SYSTEMS DESIGN AND ENGINEERING

Facilitating Multidisciplinary Development Projects

SYSTEMS DESIGN AND ENGINEERING

Facilitating Multidisciplinary Development Projects

G. Maarten Bonnema
Karel Th. Veenvliet
Jan F. Broenink

CRC Press is an imprint of the
Taylor & Francis Group, an **informa** business

CRC Press
Taylor & Francis Group
6000 Broken Sound Parkway NW, Suite 300
Boca Raton, FL 33487-2742

Printed on acid-free paper
Version Date: 20151014

International Standard Book Number-13: 978-1-4987-5126-1 (Paperback)

Visit the Taylor & Francis Web site at
http://www.taylorandfrancis.com

and the CRC Press Web site at
http://www.crcpress.com

Contents

List of Figures ix

List of Tables xi

Foreword xiii

Preface xv

About the Authors xvii

Chapter 1 Introduction 1

1.1 Development of Modern Systems 1
1.2 Systems Engineering: The Approach in This Book 1
1.3 Facilitating Multidisciplinary Projects 2
1.4 Designing a Solar Race Car 3
1.5 Notes on How to Use This Book 4

Chapter 2 The Systems Engineering Process 7

2.1 Introduction 7
2.2 The Essence of the Systems Engineering Process 7
2.3 A Practical Implementation of Systems Engineering 9
2.3.1 System Level 11
2.3.2 Subsystem Level 14
2.3.3 Integration 15
2.3.4 Verification and Validation 17
2.3.5 The Vee Model 18
2.4 A Short History of Systems Engineering 19

Chapter 3 Systems Thinking Tracks 23

3.1 Introduction 23
3.2 Dynamic Thinking 25
3.3 Feedback Thinking 27
3.4 Specific–Generic Thinking 29
3.5 Operational Thinking 30
3.6 Scales Thinking 31
3.7 Scientific Thinking 32

3.8 Decomposition–Composition Thinking....34
3.9 Hierarchical Thinking....35
3.10 Organizational Thinking....37
3.11 Life-Cycle Thinking....38
3.11.1 Product Life-Cycle Thinking....38
3.11.2 Resource Life-Cycle Thinking....40
3.11.3 Organization Life-Cycle Thinking....40
3.12 Safety Thinking....41
3.13 Risk Thinking....42
3.14 Summary....43

Chapter 4 System Design Tools....45

4.1 Introduction....45
4.2 Nine-Window Diagram....45
4.3 Context Diagram....46
4.4 Scenarios....48
4.5 Functional Modeling and Analysis....50
4.5.1 Function Trees....50
4.5.2 Functional Block Diagrams....51
4.5.3 State Transition Diagrams....52
4.5.4 Influence Diagrams....53
4.6 N^2 Diagram....55
4.7 Architectures....56
4.8 System Budgets....58
4.9 FunKey Architecting....60
4.9.1 Coupling Matrix to Budgets....62
4.9.2 FunKey as Tracking Tool....62
4.10 A3 Architecture Overviews....63
4.11 Failure Mode and Effect Analysis....66
4.11.1 Introduction....66
4.11.2 FMEA Team....66
4.11.3 FMEA Form....67
4.11.4 FMEA Procedure....67
4.12 Risk Management....71
4.12.1 Decision Tree....72
4.12.2 Risk Register....74
4.13 Documentation and Reviewing....75
4.13.1 General Documentation Guidelines....76
4.13.2 Document Contents....77
4.13.2.1 Requirements Document....77
4.13.2.2 Design Document....78
4.13.2.3 Other Documents....79
4.13.3 Reviewing Documents....79
4.13.4 Open Issues and Decisions....81

4.14 Modeling and Simulation 82
4.15 Question Generator 83

Chapter 5 The Systems Engineer at Work 85

5.1 Communication in Systems Engineering 87
5.2 The Systems Engineer and the Project Manager 89

Appendix A TRIZ 93

A.1 Short Introduction to TRIZ 93
A.1.1 Positive Priority Matrix, PM^+ 94
A.1.2 Negative Priority Matrix, PM^- 95

Appendix B Types of Failure Modes and Effect Analysis 99

Appendix C Document Template 101

Bibliography 105

Index 111

List of Figures

1.1 The World Solar Challenge across the Australian continent. 3

2.1 The System Under Design (S.U.D.) has to fit the identified issue, the context, the stakeholders and the rest of the world. 8
2.2 The Muller pyramid 9
2.3 The top-level systems engineering process as a loop of exploring the issue and defining a solution. 10
2.4 The systems engineering process..11
2.5 Subsystems and interfaces. 14
2.6 The integration process that accompanies the development process in Figure 2.4. 15
2.7 An integration step as a loop of building, adjusting and modifying versus verification. 16
2.8 Extended vee model. 18

3.1 The System Under Design (S.U.D.), the identified issue, the context, the stakeholders and the rest of the world, and one of many thinking routes (in red) for the system designer. 24
3.2 Inherent design problem. 24
3.3 The relation between availability of information and phases in development. 25
3.4 Two models for the dynamic, time-related, behavior of a solar car confronted with a puncture. 27
3.5 Basic feedback (control) system. 27
3.6 Three strategies for heating a house. 36
3.7 The ASML project structure. 38
3.8 The 2011 Solarteam Twente car...39

4.1 The nine-window diagram. 46
4.2 The nine-window diagram of the solar race car team depicting past and future features. 47
4.3 Context diagram template showing the stakeholders around a system under design (S.U.D.). 47
4.4 Context diagram for Solarteam Twente. 48
4.5 An example of a function tree. 50
4.6 An example of a functional block diagram..51
4.7 A state transition diagram for the operation of a solar racer.................52
4.8 Spiral development model. 53
4.9 Concept of a wafer scanner. 54

4.10 Logical influence diagram for a solar racer. 55
4.11 Tracking a mass budget. 60
4.12 Key drivers act as interfaces between stakeholders. 61
4.13 A3 Architecture overview of a litter-collecting robot. 64
4.14 The A3 Architecture overview creation process. 65
4.15 The severity of a risk is a combination of chance of occurrence and consequences. 72
4.16 A decision tree in its simplest form to map the possible course of a project. 73
4.17 Decision tree for analysis of bid calculations. 74
4.18 The systems engineering process shown with the reviews in place. The reviews act as feedback throughout the process. 80
4.19 The simulation process. 83

5.1 Architecture as interface between stakeholders expectations and engineering achievements 86
5.2 Architecting and monitoring the development process in the Muller pyramid shown in Figure 2.2. 87
5.3 Schramm's model of communication. 88

A.1 The TRIZ approach to design and problem solving. 94

List of Tables

1.1 Characteristics of World Solar Challenge 4

2.1 Major milestones in history seen from a systems engineer's view 20

4.1 N^2 diagram 56
4.2 N^2 diagram for a teletext to xml converter used to fill the electronic program guide (EPG) of a personal video recorder 57
4.3 A mass budget for a solar racer 59
4.4 FunKey coupling matrix $\mathscr{C}$ used to investigate contributions of functions to key drivers (kd_i) 61
4.5 The Failure Mode and Effect Analysis form 68
4.6 Severity ratings for an FMEA. 69
4.7 Occurrence and detection definitions for an FMEA 70
4.8 Evaluating the decision tree quantitatively 75
4.9 Risk register as risk inventory and management tool 76
4.10 Checklist for a requirement specification document, based on Eger et al. 2013 78

A.1 Part of the contradiction matrix used in TRIZ 94
A.2 The 40 principles identified in Altshuller [1997] 95
A.3 The positive priority matrix PM^+, found from counting the times an innovative principle is mentioned in the row of technical parameters 96
A.4 The negative priority matrix PM^-, found from counting the times an innovative principle is mentioned in the column of technical parameters 97

1 Introduction

We introduce the field of designing multidisciplinary systems with a few examples. We show that in addition to the well-known engineering fields, another discipline is needed: *systems engineering*. We will introduce it combined with the way we treat it in this book. Also, we will give an example that is used throughout the book.

1.1 DEVELOPMENT OF MODERN SYSTEMS

A digital single-lens reflex camera (DSLR camera), a medical imaging device like a magneto resonance imaging-scanner, a storm surge barrier, a modern passenger car and commercial aircraft are products that contain many parts and have to perform many functions. All parts and functions have to work together and work in an environment with users. Furthermore, some of these products have a long life in which the environment may change dramatically.

Looking inside the products, we see that there are mechanical and electronic parts, software components, styled parts, and there is behavior that is designed for good interaction with the users. Also, the products have to perform well from a financial point of view, whether that is because they are bought from hard-saved money of a consumer or paid for by tax money. At the same time the development, production and maintenance of the systems have to be financially profitable. All these aspects are dealt with mostly by specialists: electronic, mechanic, software engineers, user interface designers, product stylers, financial experts etc. Nowadays the effect is that these products, *systems* as we will call them in this book, are developed by a large team –often even several teams– of people with diverse backgrounds. Communicating and working towards a common goal is not easy in such a setting. *Systems engineering* is a discipline that aims at coping with this by focusing on the whole.

1.2 SYSTEMS ENGINEERING: THE APPROACH IN THIS BOOK

Systems engineering is an established discipline for which quite a few good books exist [Blanchard and Fabrycky, 2011; Kapurch and Rainwater, 2007; Maier and Rechtin, 2000; Sage and Armstrong, 2000, to name a few]. *Designing* systems is a less documented competence. In this book we treat a combination of the two. Although we will not go into the finest details of systems engineering, we do provide a firm basis so that the design process can take place as smoothly and efficiently as possible. It rests on three pillars:

Systems Engineering Process: How is the process of bringing systems into being organized? The process is generally called *systems engineering*.

Systems Thinking Tracks: Several ways of thinking about the system that is being designed, its context, its user, its past and its future are considered. These thinking tracks stimulate the creativity of the systems engineer, and force him or her[1] to think within and outside any limits set by the system requirements, defined in The Systems Engineering Process.

System Design Tools: These practical tools help in both The Systems Engineering Process and Systems Thinking Tracks.

We separate the three pillars, as we believe often the treatment of systems engineering is needlessly complicated by documentation processes, standards and formalities, which are in fact only itself support tools. The SE process in itself is quite simple and logical. The way of thinking in systems design, as described in Chapter 3, is a competence that is widely applicable and leads to systems that fit their purpose, not only the requirements.

Throughout the book,there will be cross-references from the systems engineering process to the thinking tracks and the tools and vice versa. These are marked in the margin. When, for instance, the N^2 diagram can be useful, it will be marked as shown in the margin here.

§4.6: N^2 Diagram

1.3 FACILITATING MULTIDISCIPLINARY PROJECTS

As for the subtitle "facilitating multidisciplinary development projects," we intend to provide a starter kit for the systems engineer in multidisciplinary development projects. One can see the systems engineer as a lubricant in this process. Please note that lubrication is more than making sure things run smoothly. Of course that alone is quite an achievement in multidisciplinary developments, and the systems engineer in cooperation with the project manager will have to ensure that. However, a lubricant also cools, reduces noise, prevents rust and seals. These activities can be seen as metaphors for the systems engineer's practice as well.

Cooling: Preventing the project from getting overheated by balancing the development load over people and time.

Noise Reduction: Are statements and reports correct or do they result in unwanted sounds? Is what is to be delivered correct or are we merely following assumptions?

Rust Prevention: The intent is to avoid decay; in this case decay of knowledge. Are the results, arguments, criteria, and code documented?

Sealing: Ensuring all parts fit together tightly. If not, the systems engineer acts as the sealing so that nothing leaks without noticing: no energy, no information, no material.

---

[1] In the remainder of this book we will refer to the systems engineer as him. This is by no means intended as a preference. It is merely to simplify the text.

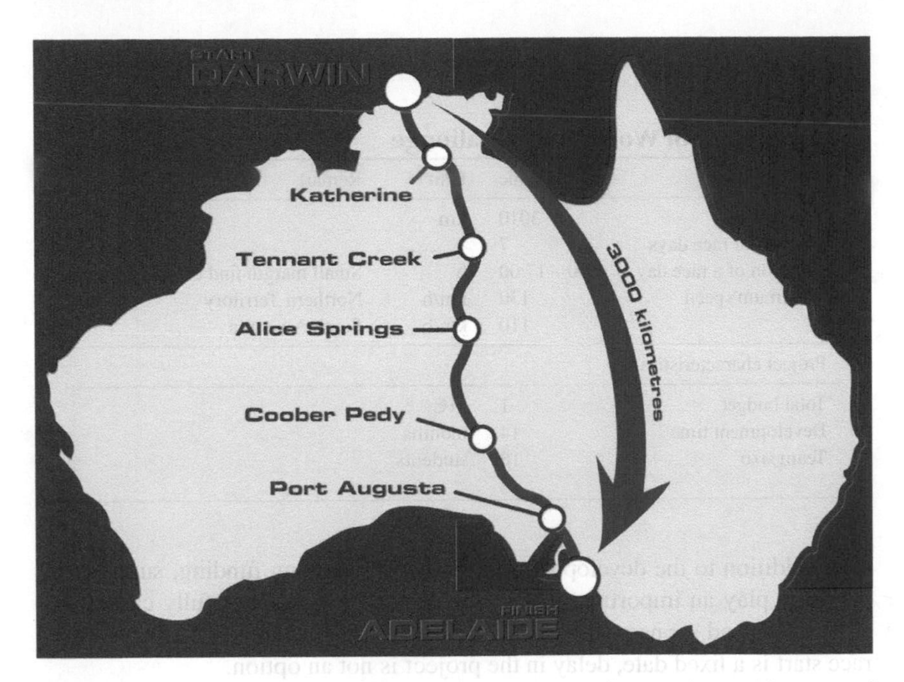

Figure 1.1 The World Solar Challenge across the Australian continent. (Courtesy of Solarteam Twente.)

1.4 DESIGNING A SOLAR RACE CAR

The book is aimed at being practically applicable. One way this is done is by being concise and to the point. Also, practical applications are used. As a running example, we use the design of a solar race car that can compete in the World Solar Challenge (`http://www.worldsolarchallenge.org/`), a race across the Australian continent, Figure 1.1. Student teams from the University of Twente competed in the challenge in 2005, 2007, 2009, 2011, and 2013. Such a development project is complicated enough to explain most of the principles in this book, yet it is not so difficult that a lot of subject information is needed before the problem can be understood. In this section a short introduction to the problem is given. The main characteristics of the race, and of the project to develop the racing car are given in Table 1.1.

The solar racer's only source of energy is the sun. The regulations have developed over the years. From 2011 onward it is permissible to have either a $3m^2$ high-performance GaAs solar panel or a $6m^2$ silicon panel. The difference in performance comes from the efficiency in conversion from the incident solar power into electricity. Where GaAs-panels can have an efficiency above 30%, silicon panels will not exceed that value (up to 25%). In addition to the solar panel, a small battery is allowed for periods with low solar irradiation. At the start of the race, this battery may be fully charged.

Table 1.1
Characteristics of World Solar Challenge

Characteristic	Value	Unit	Remark
Total length	3010	km	
Number of race days	7		
Duration of a race day	8:00 – 17:00	h	Small margin find camping spots
Maximum speed	130	km/h	Northern Territory
	110	km/h	South Australia
Project characteristics			
Total budget	1	M€	
Development time	14	months	
Team size	18	students	

In addition to the development of the car, organizing funding, support and publicity play an important role, as the project in Twente is fully carried out by students and sponsored by mainly regional organizations. Of course, as the race start is a fixed date, delay in the project is not an option.

1.5 NOTES ON HOW TO USE THIS BOOK

As stated before, this is not the ultimate guide to systems engineering. It is more an introduction tailored to get a novice in the field up and running with systems engineering. We have tried to make the material application oriented, and have condensed years of experience in systems engineering into a compact and applicable text. This book assumes a general knowledge of engineering and design although a specific discipline is not very relevant[2].

Note that it is not necessary to read the book from cover to cover, but it is highly recommended to read Chapter 2 (The Systems Engineering Process) thoroughly first, as it forms the basis for all systems engineering processes. It is more an abstract description of what the process is, and what the rationale is. Chapters 3 to 5 show how the process can be supported by systems thinking and tools, and how the systems engineer acts in his daily work.

Chapter 3 (Systems Thinking Tracks) can be read in sections, or when specific issues arise in a project. The thinking tracks that we treat are useful in a wide variety of situations. Based on the situation, the reader can select the appropriate thinking track. Alternatively, after having read the chapter completely, the reader can apply all thinking tracks in short cycles, to learn and understand the system in its context.

[2] A good reference for general design is [Eger et al., 2013], that treats the design process, introduces market research, ergonomics and creativity. It also explains systems engineering principles.

Chapter 4 (System Design Tools) can be used as a handbook or reference manual. The reader can look up a tool to get a quick start guide when the need arises.

In the last chapter covering the systems engineer at work, we discuss the way of working of the systems engineer. By means of references to the process, the thinking and the tools, we illustrate the attitude of the systems engineer. In particular, we show the importance of communication. Also, we make a few remarks about the differences between the systems engineer, and the project manager.

2 The Systems Engineering Process

In this chapter we introduce the systems engineering process: the way of working to develop large and complex systems. By investigating the problem and dividing it into smaller pieces, the work is made more manageable. The importance of identifying the interfaces between the pieces and how to put the pieces together is treated.

2.1 INTRODUCTION

Systems engineering is an accepted approach for designing large systems. In most military development projects it is compulsory to apply strict systems engineering guidelines as defined in guidelines like DOD 5000.02 [Blanchard and Fabrycky, 2011]. In the Netherlands, cooperation between government, public transport, construction companies and engineering firms has led to a guideline for systems engineering (`www.leidraadse.nl`). Using this guideline is not compulsory.

In this chapter, we present the systems engineering process concisely. One mere chapter is not enough to show all aspects of all possible situations. However, as presented in the introduction, that is not the purpose of this book. The essence of the systems engineering process in a design situation can be given in sufficient detail for an interested engineer to understand the principles of systems engineering. It is expected that with the basis given in this chapter, one can start working in a systematic way that fits the systems engineering process and apply the principles presented in the remainder of the book.

If one wants to become an expert in systems engineering and seek certification through for instance INCOSE's Certified Systems Engineering Professional (CSEP) program (`www.incose.org`), further study using, for instance, Blanchard and Fabrycky 2011; INCOSE SEH Working Group 2008 is essential.

We start with describing the essence of systems engineering in §2.2. Then, we show how this essence can be materialised in a practical way (§2.3). The chapter ends with a brief overview over the history of the systems engineering discipline (§2.4).

2.2 THE ESSENCE OF THE SYSTEMS ENGINEERING PROCESS

A system is "a set of interrelated components functioning together toward some common objective(s) or purpose(s)." [Blanchard and Fabrycky, 2011, p.17]. There are dependencies between the components in the set and between the

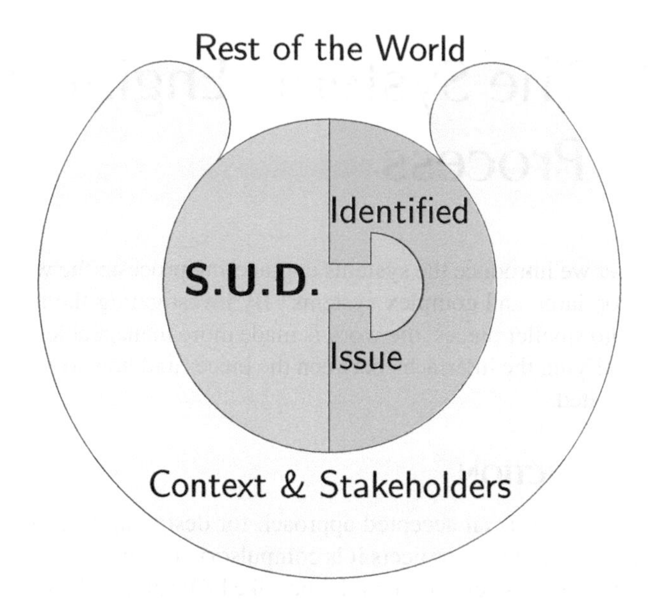

Figure 2.1 The System Under Design (S.U.D.) has to fit the identified issue, the context, the stakeholders and the rest of the world.

system in its entirety and individual components. This means that although a component can be regarded separately from the system, its full potential and its full functioning and behavior will only show in the system. Analogously, the system will only show its full potential and behavior when it is fully composed by its components and in use in a context. Moreover, the system might show behavior that was not intended, expected, or even thought possible at design time.

Figure 2.1 shows the goal of system design. A system under design (S.U.D.) is designed to deal with an identified issue. This issue can be a market opportunity, a space mission, or traffic congestion. However, the issue is not an isolated item. It is part of a context. There are often many stakeholders involved. And even when everything is inventoried, items may have been left out. (See Figure 2.1) The system under design has to fit all these issues. The systems engineering process is developed to investigate the identified issue, the context and the stakeholders, and to develop a suitable solution. It is the systems engineer's responsibility to wander beyond these inventoried issues and context to ensure proper fit between the S.U.D. and the rest of the world. The thinking tracks and some of the tools presented in Chapters 3 and 4 are particularly useful for this wandering.

§3.4: Specific-generic thinking

§3.6: Scales thinking

As the systems we are dealing with here typically consist of many components, developing is done by large to very large teams. (Please note that the principles in this book are equally applicable to development of less complex products by smaller teams.) Starting from an identified issue, the development

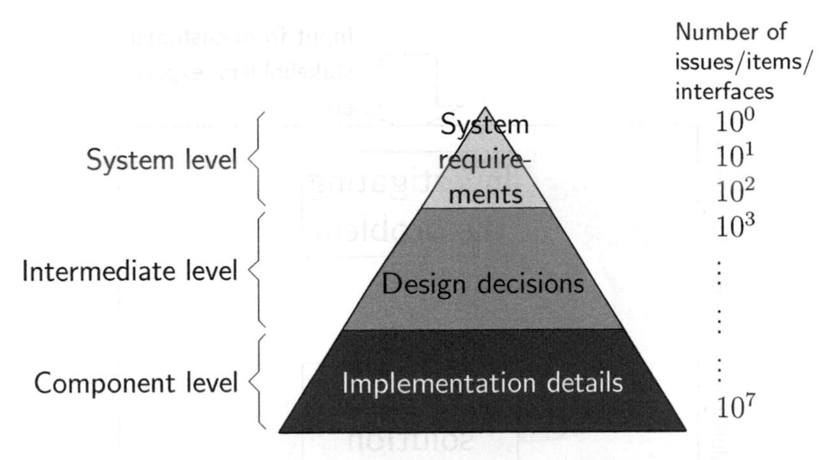

Figure 2.2 The Muller pyramid, adapted from Muller [2007]. It shows how for a complex system like an ASML wafer scanner, a Philips Healthcare MRI scanner, or a passenger aircraft, the number of issues explodes from the analysis of initial customer need to development of engineering details [Bonnema, 2011].

process deals with more but smaller details. The number of these details can be extremely large, as is shown in Figure 2.2. This pyramid shows how, starting from an identified issue with a few characteristics, the number of details explodes. A further complication is to put all these individual parts and lines of software code together—integrate them—to create a coherently working system. As a warning, that part of the development cycle is in general greatly underestimated particularly in projections of required time and effort.

§3.11: Life-cycle thinking

In the next section we explore a practical way to deal with this explosion in the number of details, and coordinating them.

2.3 A PRACTICAL IMPLEMENTATION OF SYSTEMS ENGINEERING

System engineering is, as Figure 2.1 shows, all about finding a solution to an identified issue, while considering the context, the stakeholders and the rest of the world. In fact, this is what happens at different levels of detail, see Figure 2.3: defining the issue, then finding an appropriate solution and accepting it before proceeding. Many engineers think in solutions. They even investigate a problem using a solution. With the loop in Figure 2.3 in mind this is not a problem. However, when the distinction between problem domain and solution domain is not made, one might fall into the trap of jumping to conclusions or picking the first solution that comes to mind. Every engineer, not only system engineers, needs to be continuously aware whether he is thinking in the problem domain or in the solution domain. It is important to realize that while working in the problem domain, knowledge is generated that can be used to improve the solution. In the same way, improvements to the solution may

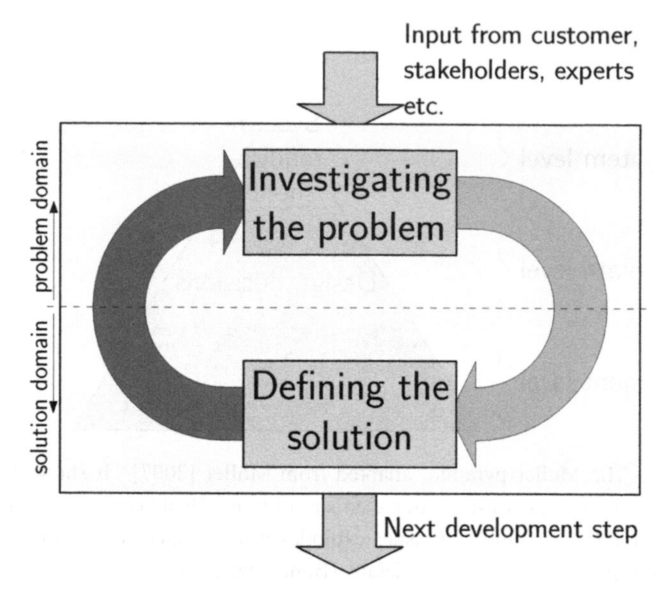

Figure 2.3 The top-level systems engineering process as a loop of exploring the issue and defining a solution. The right (downward) arrow is actually designing the solution. The left (upward) arrow depicts the verification: does the solution resolve the issue sufficiently and without too many negative effects?

reveal new knowledge about the problem. Thus it is good practice for a systems engineer to switch often between the two domains. The thinking tracks in Chapter 3 can serve as aids in this.

The solution defined from progressing the loop in Figure 2.3 one, two, or even many times describes the context for the next level of detail. It does mean that when the level of detail increases, the number of parallel processes increases. In a design process this is often seen as group of parallel sub-projects in which each of the sub-projects develops a subsystem. This is visualized in Figure 2.4. This figure shows how systems engineering repeats the same development process, shown in Figure 2.3 on different levels:

§3.10: Organizational thinking

1. System Level
   Customer Wish → Requirements → Design
2. Sub-system Level
   Requirements → Design
3. Assembly Level (depending on the field of application, this can be called, e.g., the module level)
   Requirements → Design
4. Component Level
   Requirements → Design
5. Activities

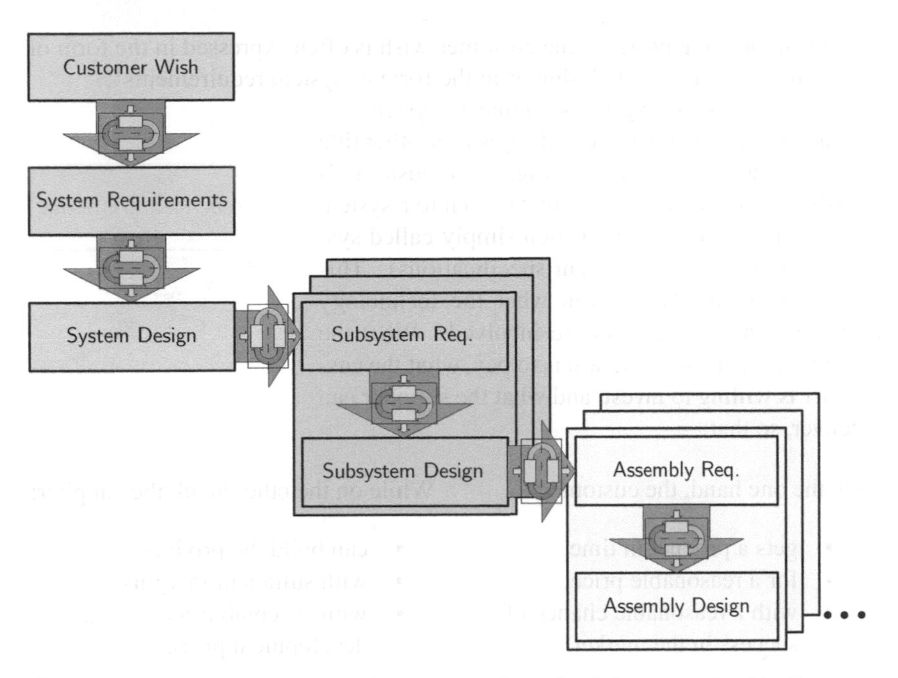

Figure 2.4 The systems engineering process. In each of the arrows in this diagram, one can see the loop from Figure 2.3.

Note that these process steps and the boxes in Figure 2.4 represent accepted baselines. Such a baseline forms an organization-wide common reference for next development steps. In Figure 2.4 it is illustrated that the work is done in the processes (arrows) between the baselines. In the next sub-sections, we treat the SE process in further detail.

2.3.1 SYSTEM LEVEL

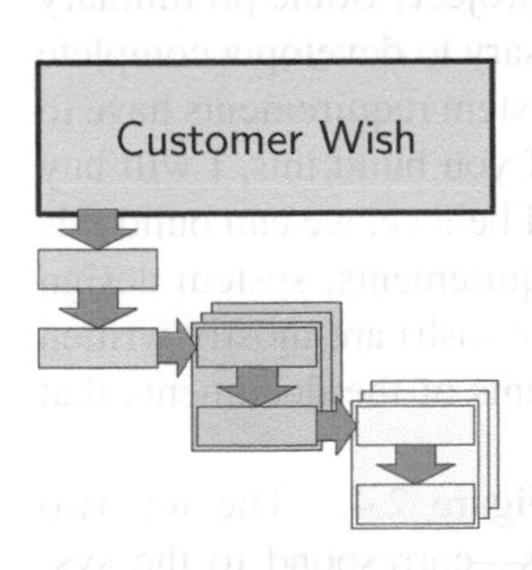

The process starts with a customer wish that may not be clearly defined. Sometimes there is a market opportunity, or a gap in the product portfolio of a company. Sometimes a customer has an idea that must be elaborated before the full development process can start. In Eger et al. 2013, this creative goal finding process is dealt with. Also, the customer need not be one person. It may be a group, or a target group defined by a marketing agency. In such cases, the marketing department of the development company expresses the customer wish. It is wise to discern between the customer of the development company, and the end user(s).The system designer has to take interests of both parties into account and balance them.

§4.3: Context diagram

§4.9: FunKey architecting

In engineering projects, the customer wish is often expressed in the form of an initial or preliminary design or in the form of system requirements[1].

The task of the systems engineer, system designer, system architect, lead designer, or other title in the organization is to investigate the customer's wish further, and to translate that wish into a system requirement specification (often simply called system requirements, or system specifications). This involves trade-offs between what the technology can offer and what risks are involved. What the end-user wants, what he wants to pay, what the customer is willing to invest, and what the supplier can deliver, so that:

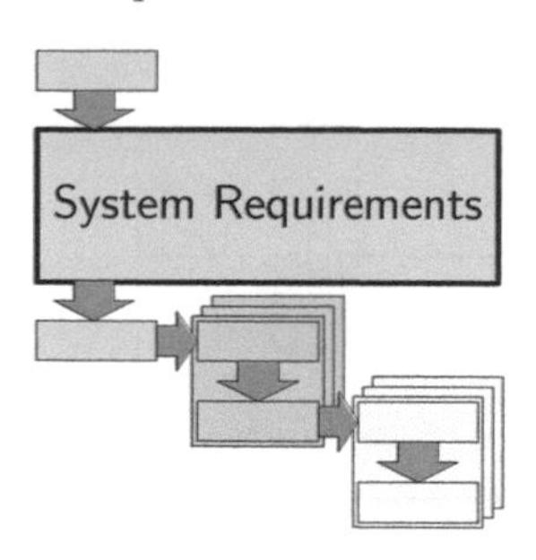

§4.12: Risk management

On the one hand, the customer:

- gets a product in time,
- for a reasonable price,
- with a reasonable chance of success in the market.

While on the other hand, the supplier:

- can build the product,
- with sufficient margin,
- with acceptable risks in the development process.

The end users will ensure success in the market if the product has a desirable combination of features (or functionality) for a competitive price. Thus, the end user's interests should be largely in line with those of the customer, if dealt with properly.

§4.9: FunKey architecting

For this, it will be necessary to not simply translate the customer's wish(es) into measurable quantities. It requires a deeper understanding of what is needed and what is possible. Most of the thinking tracks in Chapter 3 are essential for this. Further, the system designer will have to elaborate on what the customer expects from the product. At the same time the state-of-the-art technology should be known and understood. What breakthrough is imminent? What can be put to use in the time frame of this development project? Some preliminary investigation into the system design might be necessary to develop a complete and acceptable set of system requirements. These system requirements have to be reviewed and accepted by both the customer ("If you build this, I will buy it, and pay this price for it."), and the builder ("Yes, I believe, we can build this for this price and in this time frame."). System requirements, system design and all other blocks (with the exception of customer wish) are mostly written in documents. Section 4.13 contains detailed contents of the documents that are discussed here.

§4.13: Documentation and reviewing

We have looked at the left-most column in Figure 2.4. The top two blocks—Customer Wish and System Requirements—correspond to the system level in the pyramid in Figure 2.2. The system design process becomes

[1] Also called system requirement specification, or program of requirements, or similar terms.

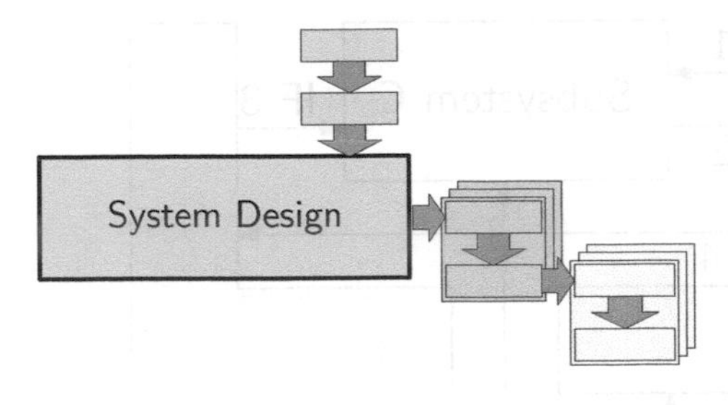

really interesting in the System Design block, and in the transition to the second column.

This corresponds with the intermediate level in the pyramid in Figure 2.2. As one can see, the number of issues increases significantly to a number that cannot be overseen easily. Also, the number of disciplines and the depth of the knowledge required from those disciplines is increasing as very diverse knowledge is required:

- What is technically feasible in the time frame before product launch?
- What is producible (e.g., can we find the right production equipment)?
- What can our development team handle (e.g., can we hire enough qualified developers)?
- What will the impact on our corporate image be?
- Will the designed system actually address the identified issue?

The thinking tracks in Chapter 3 are intended to aid in resolving these (and other) questions.

§4.7: Architectures

§4.9: FunKey architecting

§4.10: A3 Architecture overviews

§4.13: Documentation and reviewing

From the system requirements, a top-level design, or system design is created. Here, the total system is divided in its main subsystems and the interfaces between these subsystems are described. A subsystem is defined as a system that is largely independent from the other subsystems, and it performs a set of functions as described in the system design. An interface is a point of interaction between functions or components[2]. From the definition of a system in Section 2.2 it should be clear that complete independence of one subsystem from the others is impossible. If it were independent, there would be no system. *All* cooperation between the subsystems has to be made explicit by the interfaces, see Figure 2.5;

- Information
- Energy
- Materials
- Persons

§4.6: N^2 Diagram

Which of these items are transferred between the subsystems? For the systems engineering process to run smoothly, there should be no hidden transfers of any of those. All interaction takes place only via identified interfaces. Be aware that investigating the interfaces is the core business of a system designer. As noted by Robert Halligan[3], a dyed–in–the–wool system designer and trainer:

[2]In the German language, one uses the word *Schnittstelle* for interface. Literally translated, it means: “cutting place”.

[3]Founder and owner of PPI international `http://www.ppi-int.com/`

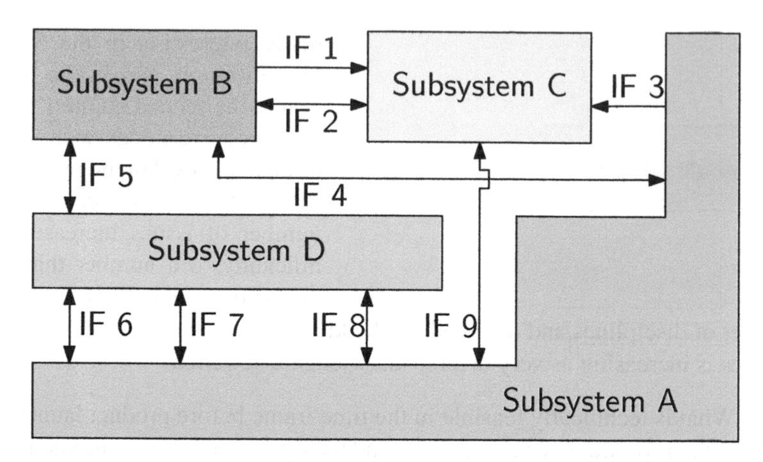

Figure 2.5 Subsystems and interfaces: all interaction between the subsystems takes place via identified and specified interfaces (IFx). Note that within a subsystem, a further splitting up can take place. Then again interfaces have to be defined.

> "There are two kind of systems engineers: those who look at the interfaces and amateurs."

While a designer is mainly interested in delivering his component or subsystem, the system designer should be focused on the things that bind these components and subsystems together: the interfaces. Therefore, the system design document specifies the performance of the subsystems, and describes the interfaces. Depending on the type of the interfaces, this can be done for instance, using CAD drawings (to ensure proper mechanical fit), signal definitions, and software class diagrams. But also conformity to standards and testing procedures should be specified as these define interfaces with the rest of the world (see Figure 2.1). The main concept choices and the design of *long-lead items* are also covered by the system design. A more detailed description is presented in the document structure in §4.13.

2.3.2 SUBSYSTEM LEVEL

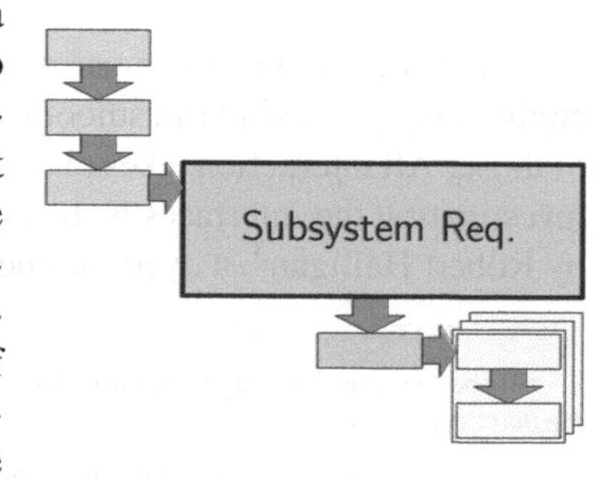

Moving to the second columns marks a change in the development project. Up to here, it was a systems-only project, involving a relatively small team. Now, in most cases, upon starting the definition of the subsystem's requirements will require assignment of the project teams for developing these subsystems. The lead designer of such a team studies and interprets the system design for his subsystem and distills the

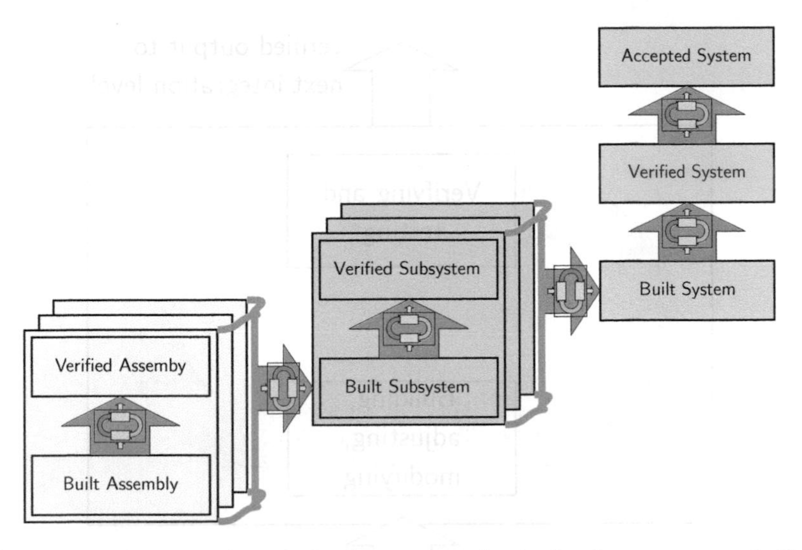

Figure 2.6 The integration process that accompanies the development process in Figure 2.4. The arrows shown are placeholders for the cyclic process in Figure 2.7.

requirements for his subsystem from it. This results in a set of subsystem requirements in much the same way as the system requirement is an interpretation of the customer wish. As there must be agreement between the customer and the supplier about the system requirement, there has to be agreement about the subsystem requirement between the systems engineer and the subsystem lead designer (and a few more people, as dealt with in §4.13).

Following the procedure used for system level, a subsystem design is created from the subsystem requirements in order to define a system that meets the requirements. The subsystem level will require more information about implementation, for example, what hardware and software will be used. §4.13 provides more information about the contents of the subsystem design document.

From this point, the process requires less system design engineering and less multidisciplinary engineering, but this depends on the scale of the initial project. It may be necessary to repeat some processes before the intermediate level shown in Figure 2.2 is reached. Figure 2.4 shows three levels (system, subsystem, and assembly), but a complex project may require six or even more levels. As the process is defined generically, at some point the system engineer can stop leading the development process and start monitoring it using hierarchical thinking and the nine-window diagram.

§3.9: Hierarchical thinking

§4.2: Nine-window diagram

2.3.3 INTEGRATION

Figures 2.3 and 2.4 suggest that developing is done by "divide and conquer". By splitting the initial problem into smaller, more manageable chunks and

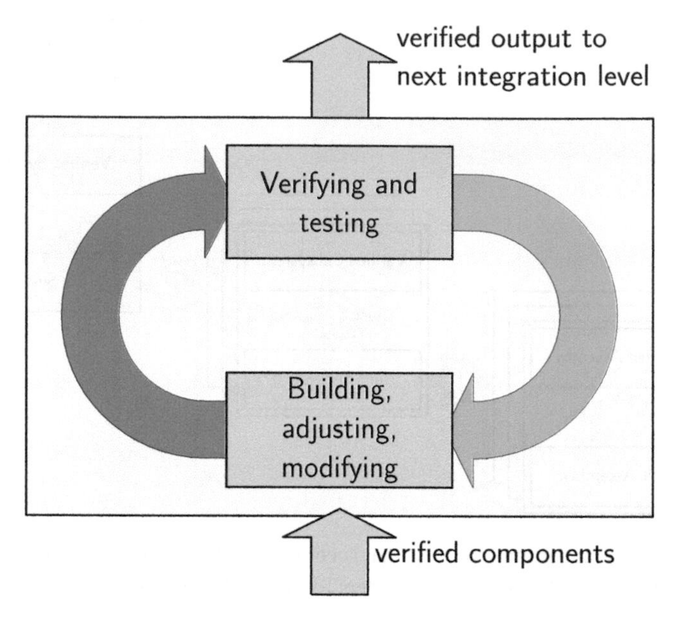

Figure 2.7 An integration step as a loop of building, adjusting and modifying versus verification. Note the similarity to Figure 2.3.

having the systems engineers watching the interfaces, the result will be a well-functioning system. This is not necessarily so, as already suggested in §2.2. When the assemblies become available, they have to be put together to become subsystems. The subsystems then have to be put together to finally form the requested system. Putting it all together and turning the power switch on will almost certainly not work the first time.

Figure 2.6 shows how integration takes place. In this case we show it starting at the assembly level, but it generally starts at component or part level. The assembly is built and verified. Although Figure 2.6 shows this as a sequential process, it is in practice a series of loops as shown in Figure 2.7.

The verification parts in Figure 2.6 and 2.7 ensure that the individual components/assemblies/subsystems are in agreement with their requirements. Thus prior to proceeding to the next step in the integration process, verification (by means of testing, simulation, etc.) of the individual components/assemblies/subsystems has to be done (see §2.3.4). The outcome of the verification can be

1. success,
2. failure,
3. minor deviation that needs fixing, but is no barrier for further integration.

Whatever the outcome, it has to be documented carefully! Engineers tend to trust their work (of course) and thus expect success of the verification; they make their plans accordingly. However, in practice many issues occur during the verification and further integration processes. Therefore, significant of time has to be reserved for integration and verification! According to Muller 2011a, different types of problems are encountered, when integrating a system or subsystem. The typical order is:

§3.8: Decomposition–composition thinking

1. The (sub)system does not build/cannot be made/does not fit.
2. The (sub)system does not function.
3. Interface errors.
4. The (sub)system is too slow.
5. The (sub)system does not meet main performance parameter(s).
6. Reliability is not met.

Clearly, issues 1–2 are on the lower level in Figure 2.7, while the issues 4–6 are on the upper level in Figure 2.7. The interface errors (issue 3) refer to both levels, and even link them.

The way to organize the integration process is to start working with hardware/software set-ups as soon as possible. Even when the set-ups contain mock-ups and simulated components, learning and knowledge creation are very important, and will result in saved time later in the process when the real material is on site. Present-day approaches with hardware, or software in the loop are aimed at reducing the integration process by taking them into account from the earliest development steps.

2.3.4 VERIFICATION AND VALIDATION

A very important aspect of the system engineering process is to verify a proposed solution before proceeding with further development. There are several verification methods applicable:

- Reviews
- Experiments, for instance by creating functional models
- Comparison to existing products
- Analyses and/or simulations

§4.13: Documentation and reviewing

§3.7: Scientific thinking

Verification is a general term for checking performance against requirements using any of the methods listed above and the ultimate goal is to deliver a system to the customer, and have it accepted by him because it solves his problem. In other words, the S.U.D. has to create *value* for the customer. Therefore, we use the term *validation* for a specific form of verification, namely verification against the customer needs and wishes, and investigating in what way the system can generate value. In systems engineering literature, and legal literature, the goal is to deliver the system that was requested and agreed upon. While that will always remain formally the maximum achievable, the goal of the systems engineer should be to deliver the system that solves the problem and generates

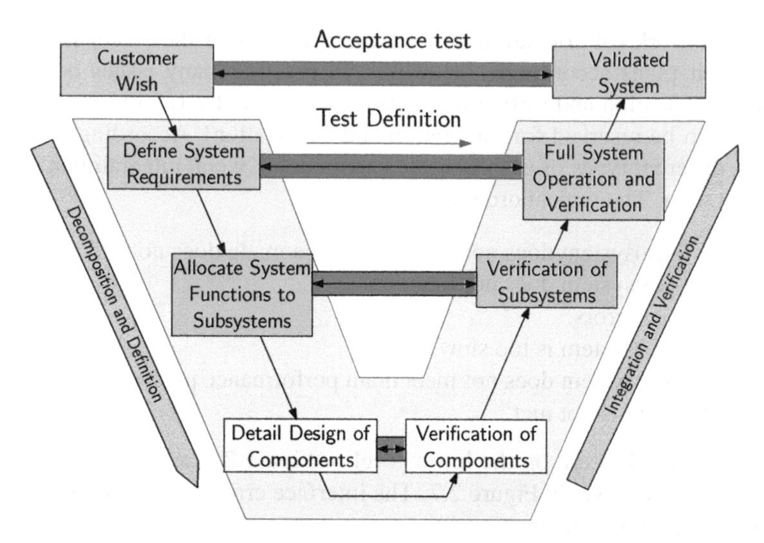

Figure 2.8 Extended vee model (adapted from [Blanchard and Fabrycky, 2011]) showing decomposition of the system into subsystems and components and the subsequent building of the system. Also shown are the definition of tests, the customer wish and the validated system.

value. Thorough understanding of the customer processes and business model is therefore essential for the systems engineer. Without it, trade-offs made by the systems engineer are likely to result in sub-optimal solutions for the customer. The descriptions of verification and validation above may differ from ones used elsewhere. Those two terms results in much clarity.

The accepted system conforms to a test[4] that is agreed upon between the customer and system builder. There should be close correspondence between the system requirements and these test specifications.

Note that a possible outcome of any verification and validation step can be a new development cycle, as depicted in Figures 2.3 and 2.7.

2.3.5 THE VEE MODEL

§3.8: Decomposition–composition thinking

§4.13: Documentation and reviewing

The vee model that is widely used in systems engineering literature Blanchard and Fabrycky 2011, tries to integrate the development and integration work. Figure 2.8 shows three levels of the system decomposition. In the extended vee model we present here, validation is made explicit by showing the customer wish and the accepted system, on top of the "traditional" vee model. This model shows that while the systems requirements are compiled, there has to be a testing procedure devised to show the specifications are met when the full system is built. Thus these testing and verification procedures have to be

[4] Customer acceptance test.

compiled in conjunction with defining the requirements, not when the development process is already well underway. Also, the requirements should be test*able*.

Also on the next level, the subsystems have to be verified according to their respective subsystem requirements. Note therefore the corresponding colors between Figures 2.4 and 2.8.

As for the definition of testing procedures: If a requirement cannot be verified by a testing, measurement or other evaluation method, its value as a requirement should be seriously questioned.

§3.3: Feedback thinking

An important note is that the vee model is not so much an organizational model; it is more a reference model that orders the knowledge created in developing systems.

2.4 A SHORT HISTORY OF SYSTEMS ENGINEERING

Effective systems engineering starts by mutual consultation and agreement between disciplines about the difference between defining *what* must be done and *how well* it must be done to determine what *should be*, what *can be* and what *is* (see Figure 2.3). To come to this effective systems engineering approach two elements should be added to a project that are traditionally not present [Arunski et al., 1999]:

- A disciplined focus on the end product, its enabling products, and its internal and external operational environment (i.e., a system view)
- A consistent vision of stakeholders' expectations independent of daily project demands (i.e., the system's purpose)

This approach developed over centuries. There is no particular date linked to the origin of systems engineering. From the past the heritage of systems engineering can be considered in complex projects and specific ways of working. Noah's Ark was built to a system specification. Other examples from history are shown in Table 2.1.

The main turning point in the maturity of systems engineering as a distinct engineering discipline arose from the increased and complex demands in the fields of military, missile and aerospace during and after the Second World War. Solving complex problems in very short times with high quality and low costs needed advanced technologies and new approaches in planning, technical collaboration and management processes. New revolutionary advances in transportation and application of materials, information and energy were needed that could only be developed by thinking "out of the box" and over the boundaries of existing disciplines like electrical, mechanical, civil, chemical engineering and physics and mathematics [Kossiakoff and Sweet, 2003]. The cooperation between the systems engineering approach and traditional engineering approaches became more mature, strengthened by the military demand and space and aircraft engineering during the Cold War of the 1950s, 1960s

Table 2.1
Major milestones in history seen from a systems engineer's view

4000BC	Water distribution systems in Mesopotamia
3300BC	Irrigation systems in Egypt
400BC	Development of urban systems as in Athens, Greece
300BC	Construction of Roman highway systems
1808–1825	Design and build of water systems like the Erie Canal
1877	Development of telephone systems
1880	Electrical power distribution systems
1937	British multi-disciplinary analysis of air defence system
1939–1945	Supporting Nike missile development by Bell Laboratories
1948–1967	RAND corporation developed operational research and systems analysis approaches
1951–1980	SAGE air defence system defined and managed by MIT
1954–1964	ATLAS intercontinental ballistic missile program managed by Ramo-Woodridge Corporation
1964	Nine universities offered systems engineering programs
1966	USAF *Systems Engineering Handbook*
1969	Systems engineering management MIL-STD-499 (USAF)
1979	U.S. Army Field Manual 770-78, systems engineering
1983, 1990	Defense systems management college and systems engineering management guides
1990	Formation of the National Council on Systems Engineering (NCOSE)
1995	NASA systems engineering handbook
1995	NCOSE changes to the International Council on Systems Engineering (INCOSE)
1998	International Council On Systems Engineering (INCOSE) *Systems Engineering Handbook* version 1 appeared
2001	Federal Highway Administration's (FHWA) handbook title "System Engineering for Intelligent Transportation Systems"
2005	First release of SysML, to be used in model-based systems Engineering (MBSE).
2009	CNNMoney rates systems engineer "Best Job in America"
2010	SysML1.0 language certified `http://www.omgsysml.org`
2012	First release of the SEBoK: Systems Engineering Body of Knowledge (`sebokwiki.org`).

and 1970s. A major driver in the development of systems engineering and other engineering approaches became the "information age". Development of the digital computer in combination with advanced software technology gave a boost in supporting problem solving by engineering teams in simulating the performances of complex systems as a whole in relation to the necessary human interactions before the systems were built. New and innovative products, services and processes arose in a very short period and in this flow of technology and management modern systems engineering grew to higher capability levels.

The relations of systems engineering to its origins can be understood in terms of three basic factors (all cited from Kossiakoff and Sweet 2003):

1. Advancing technology, which provides opportunities for increasing system capabilities, but introduces development risks that require systems engineering management
2. Competition, whose various forms require superior (and more advanced) system solutions through the use of system-level trade-offs among alternative approaches
3. Specialization, which requires the partitioning of the system into building blocks corresponding to specific product types that can be designed and built by specialists, and strict management of their interfaces and interactions.

3 Systems Thinking Tracks

To achieve the optimal fit between the system under design (S.U.D.), the identified issue, the context, stakeholders and the rest of the world, a system designer has to think along several lines. In this chapter we elaborate a number of these lines based on scientific research, literature and experience from the authors.

3.1 INTRODUCTION

This chapter shows another side of systems engineering. It connects more to the *design* part of the title of this book. As design is not finding the answer to a closed problem where there is only one solution, we have to look for inspiration on the one hand and constantly evaluate possible solutions on the other hand (see Figure 2.3). In literature several approaches are presented. Some are more in the line of creativity [De Carvalho and Back, 2000; Eger et al., 2013; Gelb, 1998], others use lines of reasoning or systems thinking tracks [Boardman and Sauser, 2008; Muller, 2011b, 2004a; Richmond, 1993]. Finally, there are descriptions of how to look at systems [Boardman et al., 2009; Martin, 2004]. Combining these with experience from the authors, we show in this chapter a series of systems thinking tracks that can be used as:

- A checklist
- A viewpoint changer to avoid mental inertia and improve the fit between system and context
- A creativity starter

It is important to realize that the procedure in Chapter 2 is rather precise and results in a description of the system to be designed that will fit the identified issue, the context and the stakeholders. However, how it will fit the unknown rest of the world is not treated. The thinking tracks given here[1] will deliberately pull the thinking of the system designer out of the identified issue–S.U.D.–context–stakeholder path, as shown by the more or less chaotic red path in Figure 3.1.

An inherent problem in *design* is shown in Figure 3.2. At the start of the design process, the outcome is unknown, while the freedom to make decisions is infinite: the controversy of system design. In that situation, the architecture of the system has to be defined (see §4.7), which shapes the overall system. This is a typical situation that describes the core of the system designer's job: making important and far-reaching decisions, based on inadequate and uncertain information. The information that *is* there, should be made as accessible

[1]Sections 3.2 to 3.7 were inspired by and generalized from Richmond [1993].

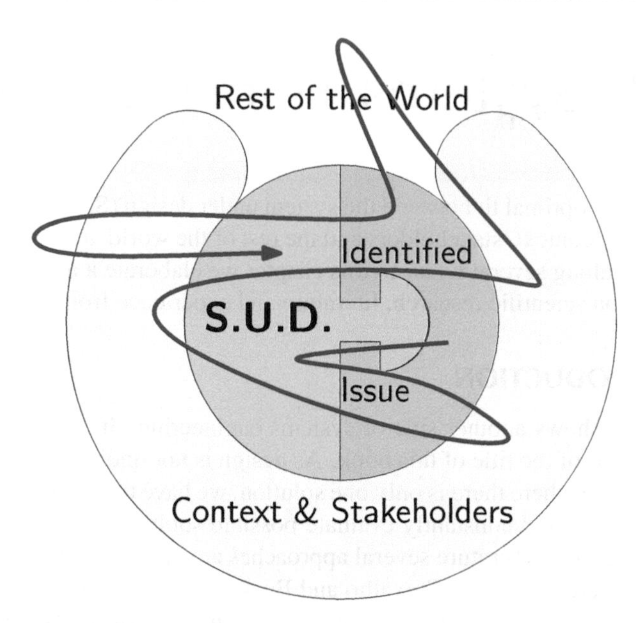

Figure 3.1 The System Under Design (S.U.D.), the identified issue, the context, the stakeholders and the rest of the world, and one of many thinking routes (in red) for the system designer.

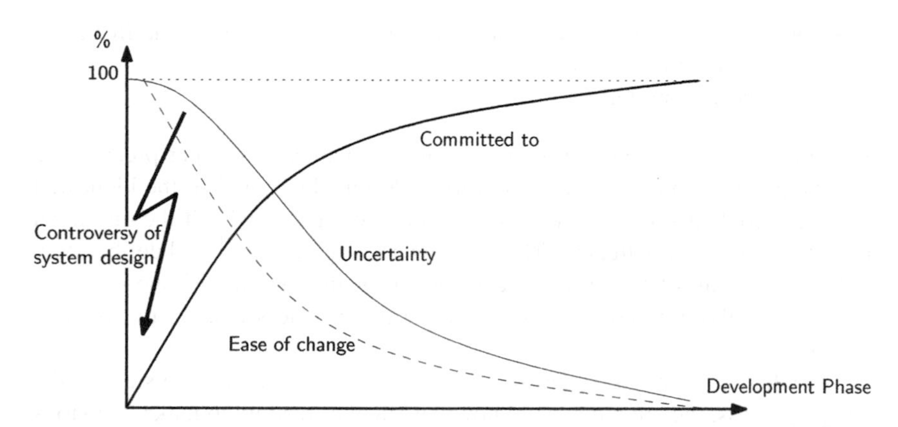

Figure 3.2 Inherent design problem. In the early phase of development, many decisions have to be made, and technologies have to be committed to significant uncertainty (or lack of knowledge) that exists. The ease of change decreases rapidly as a project progresses. (Adapted from Blanchard and Fabrycky [2011].)

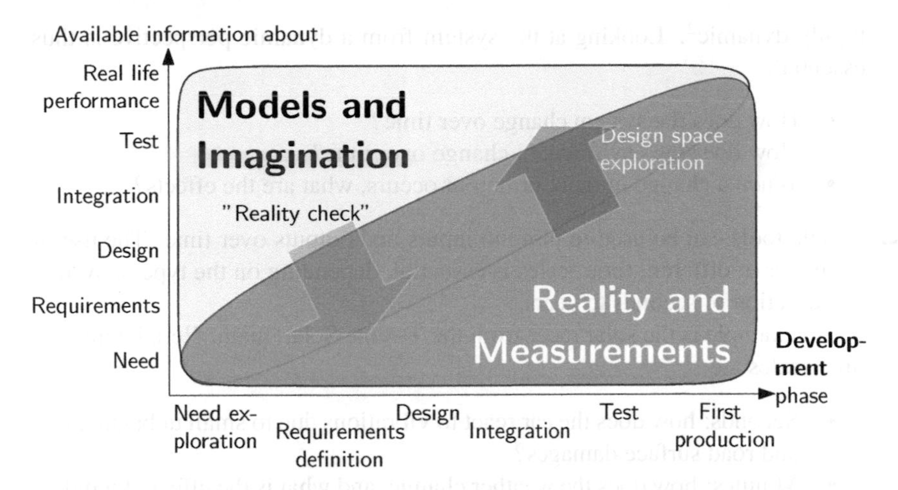

Figure 3.3 The relation between availability of information and phases in development. The use of models and imagination complements reality and measurements.

as possible. The situation is even more complex because the system itself will impact the issue, context, the stakeholders and the rest of the world. This is one of the most discerning aspects of systems design compared to product design.

The thinking tracks presented here and most of the tools in Chapter 4 are meant for dealing with this uncertainty. Yet, one should be aware that to remove most of the uncertainty, the whole design process has to be completed. This, of course, is not feasible. Therefore, the system designer has to sample the whole design field, using the thinking tracks and tools presented in this chapter and the next. Only the essential and risky parts will be given more attention in the early part of the process. The other issues can be left to a later stage, other than specifying their performance and interfaces (see §2.3). Determining what belongs to which category is largely a matter of experience.

The thinking tracks discussed in this chapter, and the tools of the next chapter all help in uncovering information and knowledge or sometimes only making the knowledge explicit and (more) readily available. This requires often a combination of modeling, measurements, reality checks and imagination, as shown in Figure 3.3. There is a contrast with Chapter 2 in which a process was described that can be followed. Here, the *softer side* (as opposed to formal side) of systems engineering is shown. This less formal side is in the opinion of the authors not less important. The use of examples and metaphors illustrates this.

3.2 DYNAMIC THINKING

The systems designed by system designers and systems engineers are not monolithic invariant systems. They interact with the environment and are often highly dynamic[2]. Looking at the system from a dynamic perspective is thus essential:

- How does the system change over time?
- How does the environment change over time?
- When a change in input or output occurs, what are the effects?

§4.14: Modeling and simulation

Simple tools can be used to plot the inputs and outputs over time. The use of a number of different time scales is essential, depending on the type of system and duration of seconds to years.

An example is the solar racer from the Twente Solar Team. Here interesting time scales are:

- Seconds: how does the car react to vibrations due to small unbalances and road surface damages?
- Minutes: how does the weather change, and what is the effect of wind gusts? How will it react to a puncture?
- Hours: driver behavior and feeding and short-term strategy;
- Days: overall strategy and race planning, including foraging, camp planning vehicle manning etc.
- Weeks: project planning and manufacturing, communication strategy.
- Months: financial planning, motivation, training and overall project planning.

The systems engineer can model the impact of a puncture on the total racing time. This can be done for velocity as function of time; see Figure 3.4(a). From this, conclusions can be made about the time lost. However, what consequences this time lost imposes on position in the race is hard to predict. Therefore an alternate model, or merely another representation of data already present in the model, may be needed, namely distance as function of time; see Figure 3.4(b). Now, the trajectory of a competing car that has no puncture can be easily incorporated; see the dotted line. From these two diagrams design consequences can be derived: changing a tyre has to be done fast, but also accelerating after a puncture should be fast. Maybe a device to store electricity close to the motor can help here. Another modeling switch that has to be made often in dynamic systems is that between time domain and frequency domain. The two models in Figure 3.4 are both time-domain models. In particular when the time period of interest decreases, it becomes more interesting to switch to frequency-domain models and diagrams like *Bode plots*. Many systems react differently to slow disturbances and fast disturbances.

Fortunately, modern modeling and simulation packages can easily output different diagrams like the two in Figure 3.4; also they can output frequency responses and transfer functions; see §4.14.

---

[2]Definition of dynamic: always active or changing (Merriam-Webster online, retrieved 20140320).

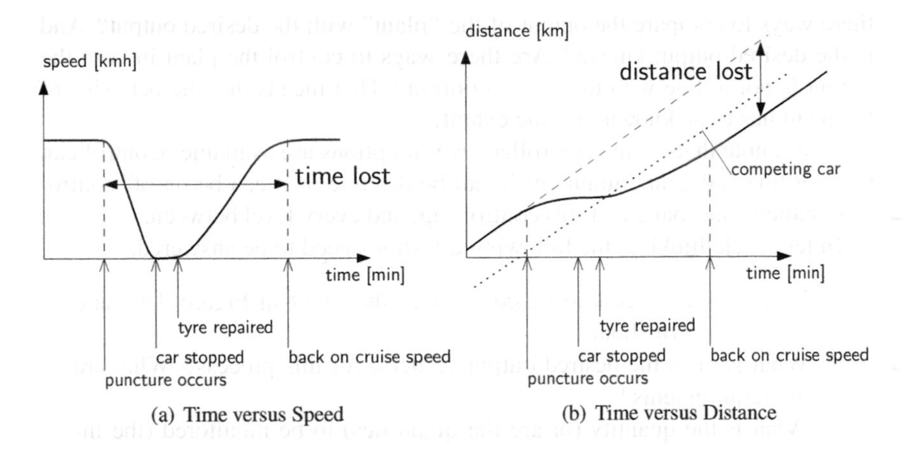

(a) Time versus Speed (b) Time versus Distance

Figure 3.4 Two models for the dynamic, time-related, behavior of a solar car when confronted with a puncture. In 3.4(a) the time lost is shown; in 3.4(b) the position relative to a competing car is shown.

3.3 FEEDBACK THINKING

Closely related to the fact that systems have time-related behavior is the question whether there is feedback. The principle of feedback is comparing the measured output of a system with the desired output, and adjusting the available controls so that the requested output is achieved. Figure 3.5 shows a basic feedback (control) system: something that has to be controlled, traditionally called the *plant*, creates an output. The output is measured by a measurement system. The result is compared to the desired output (the setpoint), which is given as an input to the complete feedback control system. The error, ε is used in a controller to create control signal inputs for the plant to create the desired output.

There is a large body of knowledge on feedback control systems that will not be repeated here (see for instance [van Amerongen 2010] for an introduction). When designing systems it is important to look for feedback mechanisms. Are

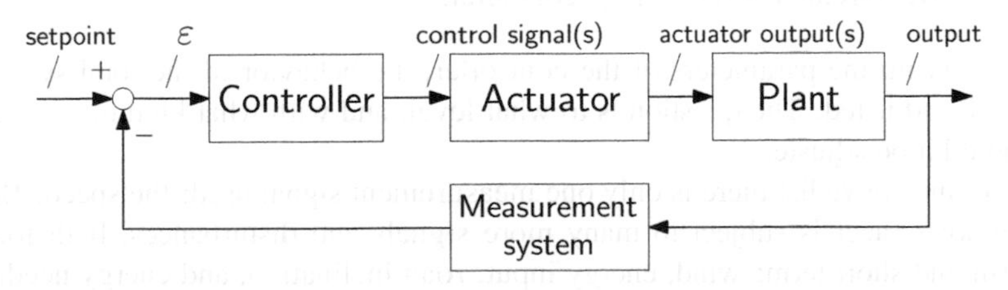

Figure 3.5 Basic feedback (control) system. The "plant" is the component to be controlled.

there ways to compare the output of the "plant" with the desired output? And is the desired output known? Are there ways to control the plant in case the output is not in line with the desired output? That means that the behavior of the plant has to be known to some extent.

Please note that for the controller several options are available. Control can be very fine and gentle tuning or it can be discrete or it can be on-off control (also called bang-bang control) controlling, and every level between.

§4.14: Modeling and simulation

In feedback thinking, the following questions need to be answered:

§4.13: Documentation and reviewing

1. What is the process to be controlled (the *plant* in Figure 3.5), and what is known about it?
2. What is (are) the desired output value(s) for this process? What are the requirements?
3. What is the quantity (or are the quantities) to be monitored (the input(s) to the measurement system in Figure 3.5)?
4. How do we measure this quantity (these quantities)? Is a sufficiently accurate *measurement system* feasible? Is the response time of the measurement system sufficiently short compared to the expected changes?
5. Is the *plant* controllable? Are there ways to control the plant, so that the output of the plant changes predictably?
6. Can a *controller* be devised? To answer this, study of control engineering may be necessary [van Amerongen, 2010; Bolton, 2012; Onwubolu, 2005].

Modeling and simulation are useful and often even necessary tools for these kinds of problems.

In the solar racer, the speed is controlled using a cruise control system. In its basic form, the system consists of the following; see Figure 3.5:

- Controller: a piece of software that works using control algorithms.
- Actuator: the electric motor including its power amplifier that drives the car. The actuator is steered by the controller.
- Plant: the solar racer that has to move as fast as possible. The car is affected by friction, road inclination and other disturbances.
- Measurement system: a speedometer.

By varying the parameters of the controller, the behavior of the total system can be adjusted. The question is to what level, and with what kind of strategy should it be adjusted?

§3.9: Hierarchical thinking

In the above list there is only one measurement signal used: the speed. But the solar racer is subject to many more signals and disturbances, both long term and short term: wind, energy input, road inclination, and energy needed for overtaking vehicles. If these can be measured or predicted and the controller can be taught to cope with that, the overall behavior can be optimized to parameters like speed and energy use.

When disturbances are known in advance, one could extend the feedback control system to also include *feed-forward control*. Using a model of the plant and actuator, additional control signals are generated based on the disturbances and desired behavior.

§3.10: Organizational thinking

Feedback thinking does not apply to the SUD alone, but also to the organization that is set up to create the system. It is a valuable way of looking at the project organization. By giving frequent feedback about the current state of affairs relating to planning (how much delay, money spent), the context (e.g. the company in terms of financial status, people status, plans etc.), and the rest of the world (e.g., the market), people will understand their position in the big picture of the whole organization, and know that their contribution is essential. Tools like lean manufacturing and knowledge-based production are very suitable on the project level [Mascitelli, 2011]; we will not treat these in this book.

3.4 SPECIFIC–GENERIC THINKING

§4.9: FunKey architecting

When discussing how to undertake a business trip to another part of the country many motorists use the argument that "public transport is always delayed, and crowded". The crowded claim may be true during rush hour, but the always delayed sitution is often based on one or two incidents from the past. If these specific situations are then used as generic facts for every transportation situation, wrong decisions are imminent.[3]

Specific-Generic thinking is about the scale of the problem and of the solution. Is it worth devising an automated system to level a painting on a wall, when doing it by hand once a month will do? In politics (but not only in politics) this often occurs: the government minister is called to answer questions about an incident, while 99.9% of all situations are handled very well. Then the parliament is so shocked that a new committee is installed to investigate and organize this exception and all 99.9% normal cases. This results in increased numbers of government employees and increased government spending for which the minister is called to the parliament again.

Although the solar racer has to drive on public roads, and thus has to show its road worthiness, it does not have to comply with all rules and regulations of the Road Traffic Act, but only to the basic ones. Two accompanying vehicles that fully comply drive in front of and directly behind the solar racer. This specific solution to a problem enables developing a solar racer by students within the limited time frame. The alternative of requiring complete compliance with the Road Traffic Act (the generic solution) for racing on the Stuart Highway on solar energy (the specific situation) requires much more design work.

There is a link with scientific thinking (§3.7) in the sense that one measurement cannot be used to prove something. Yet, there is an inherent problem here

[3] A recent study confirmed this overestimation of public transport traveling time by motorists, [van Exel, 2011].

in design, as illustrated in Figures 3.2 and 3.3. In the early phases of the design project, everything is unknown. Nearly all decisions have to be made based on models, estimations and even imagination. Measurements can only serve as a "reality check" for the order of magnitude as they are performed on systems that may behave very differently. As development progresses, more knowledge becomes available, so less imagination is needed and the measurements that can be done are more relevant for the system under design.

§4.8: System Budgets

In this track it is necessary to determine the scale of magnitude of the problem, and the estimated investment for the proposed solution. The moment numbers are introduced in the discussion –even if they are off by a factor of two– will already filter out many "shooting a mosquito with a canon" situations.

3.5 OPERATIONAL THINKING

In operational thinking it is investigated how something is done in the real world. This is best explained using an example: our solar race development. As shown in table 1.1, a race day starts at 8 o'clock: from that moment onward it is allowed for the car to drive. Operational thinking is then analyzing what is necessary to come to this point: what operations and procedures have to be undertaken to get the car (and all other vehicles) up and running safely and in time, so that no valuable race time, nor valuable energy are wasted?

Issues to consider for this specific case are (among others):

- Waking up, making and eating breakfast
- Aligning the solar panel with the sun the moment the sun rises
- Starting the solar car's systems
- Technical check of the solar car
- Updating all model parameters (weather, competitors, etc.)
- Sending press updates
- Packing the cars and setting up the convoy
- Taking down the tents and cleaning the area
- Health and safety checks

A similar list can be made for all stops and for setting up the camp at the end of the day. Exceptions have to be considered as well. One can think of dealing with punctures, damages, accidents, bad weather, and other issues. For other systems, other lists have to be made. As this thinking track relates to how the actual system works, it is not possible to make a generic list that works for all systems. However, there are a few guidelines that can be applied:

§4.4: Scenarios

- Compile a story, storyboard or scenario of the whole process.
- Include the start-up and shut-down phases, or even write specific stories for these phases. In many cases these are complicated, whereas the normal operation phases are relatively simple.
- Investigate exceptions: emergency stops, failures and other incidents.

Makes these stories in a form presentable to experts, and to people who have experience with operating comparable systems. Meet and talk with them, or even better, involve the operators in this process. There is a wealth of practical information available from the operators. System Engineers sometimes have a fear going to the operators. Overcome that fear.

3.6 SCALES THINKING

Scales thinking is about finding the nuances in arguments, working with uncertain and incomplete information and about avoiding creating opposing camps.

Engineers are educated to work with exact and verifiable data. As mentioned before, in design, uncertainty plays a large role: the problem is not exactly clear (although it may be clearly written down and reviewed, there may be a mismatch with the actual problem; see Figure 3.1), nor is the outcome known at the beginning. In design, options are created that are compared and choices are made based on arguments and data. However, arguments and data may contradict. It is a rare situation that one option outperforms the others on all aspects.[4]

Computer simulations are regularly based on linear or linearized equations. In reality, systems behave linearly only on a part of the spectrum. Even more so, when a technology is stretched, often a paradigm shift has to be made. One example is the telephone. Originally, the connections were made by hand, by a telephone operator. With the increasing number of connections this became impossible, so a paradigm shift was needed: automated telephone switches. At first these were electro-mechanical switches. Nowadays, these are completely digital. The shifts from manual to electro-mechanical, from electro-mechanical to fully electrical, and from electrical to digital meant changes in relation between for instance cost and performance; or performance and energy consumption; or the space and number of lines they can serve. These shifts cannot be easily modeled and are beyond linearized models. In this case, one can see that a technology has a limited scale of application. The same holds for measurement systems, actuators, and even managers! If one would have plotted the number of lines against the number of operators, then extrapolation (and thus expectation) of manual switching would have excluded common use of the telephone. The paradigm shifts that came with electro-mechanical automation, electric automation and digitization (and the corresponding scale changes) have made telephony a world-wide commodity.

For each shift, new scales of measures hold. Also, new limitations and opportunities exist. In the case of a new and unknown technology, these limitations and opportunities have to be considered with care. When they involve treating humans or when the food chain is affected, rigorous measures have to

[4] If that *is* the case, one should thoroughly question the seriousness of the other options and the effort spent in creating these alternatives.

be taken. An example from history, is the (accidental) introduction of rabbits in Australia. Because of the lack of natural enemies, the rabbits spread fast, and became a plague. Even to this day, controlling the rabbit population in Australia is an issue.

Another way of using scales thinking is the difference between yes and no or black and white and shades of gray. For instance, many people read the date on a carton of milk as the absolute date after which it cannot be consumed. However, it is a statistical prediction based on the process of producing and transporting the milk. This includes the average time the consumer leaves the milk outside the refrigerator after purchase. When someone only takes the carton out to pour a glass of milk, and maintain refrigeration at 4°C, then the milk is well drinkable one or two days after the date on the carton. However, when someone leaves the milk on a table in the sun for two hours, the milk will sour quickly. So what appears to be a clear division may be less clear when analyzed to more detail.

This way of thinking can be put to use for the system thinker in two ways. A stuck argument, that is in two (or three) camps, can be resolved by analyzing the grey scales between the two camps. On the other hand, when there is too much nuance (or uncertainty), one can find the pain barrier by increasing the limits further in a thought experiment[5] (SE: systems engineer, E: expert):

SE: How much power will you need for this function?
E: I have no idea.
SE: Oh. Well, can you make an estimate?
E: No, really, I don't know.
SE: Hmm, shall I put, 500 W then?
E: No, that is way too much.
SE: Right, is it more like 50W?
E: Well, 50 W may be a bit too little.
SE: What power do you think then. Or can you give a range?
E: It will be somewhere between 75 and 100 W.

Of course, such an experiment does not give an exact number, but at least it yields a clearer value. In subsequent improvements at later stages, the actual number can be found. Of course, additional measurements and other experiments can be useful too. This relates to the next thinking track: scientific thinking

3.7 SCIENTIFIC THINKING

Maybe the last part of the previous section gave a different impression, but in systems engineering and systems thinking, it is important to use scientific principles:

---

[5]Representation of a real-life conversation.

1. Formulate a question.
2. Formulate a hypothesis that can include a theory that explains the phenomenon observed.
3. Create a model that can be used to predict behaviour.
4. Verify the hypothesis and thus the model, by means of:
   - Simulations.
   - Experiments.
   - Consulting literature.
   - Consulting experts.
5. Analyze the results and adjust the model if necessary.
6. Adjust the theory if necessary.

The verification step is *essential* in this process. The model can consist of or be based on many different things. Of course, a computer model can be used, but also a previous version of the system, or a comparable system can be taken as a model. One can build a *functional model* that performs using the same principles, but will not be used as a production type. One can use experts' opinions as models (then use several experts). Make sure the results are verifiable, and take into account that measurements have limited accuracy.

A short, yet useful, introduction to the use of theory, models and measuring in science and engineering is presented in the first chapter of [Giancoli 2008]. The remainder of this reference is also a good source for knowledge that can be used in scientific, specific-generic and operational thinking. One particular issue is that increasing the number of measurements improves the accuracy (by reducing the error margin with a factor of $1/\sqrt{n}$ for n independent measurements). However, a systematic error cannot be reduced by repeated measurements. Also, consider that, as noted in §3.6, measurement systems also have scales of application. At the edges of their applicable range, deviations from the expected behavior may occur. Consulting the calibration data or making calibration tables can be useful.

In the development of the solar racer, the total drag coefficient is determined by roll resistance and air drag, plus internal resistance. The former two are at the beginning unknown as the number of wheels, their size, the shape of the body, etc. are open. Information is created during the project, for instance; data from preceding years can be used and simulations are run. As soon as a mock-up is ready (Figure 3.8(b)) data on roll resistance can be measured. Air drag can be measured with a model in a wind tunnel. As soon as the body is ready, real drag resistance data can be acquired at different speeds and different angles to the wind. This does require a well-thought out measurement protocol.

As for more information on scientific research in the field of systems engineering where case studies are often used, [see Blessing and Chakrabarti 2009; Martin and Davidz 2007; Muller 2013].

3.8 DECOMPOSITION–COMPOSITION THINKING

In many cases, systems engineering is presented as a way to determine how the system can be decomposed in subsystems, assemblies, parts, and components. What often is left untreated in education is how can all these be integrated into a well-functioning system. And how can, in the whole process, the integrity be safeguarded? Decomposition–composition thinking may be the thinking track that has to be practiced most frequently by the systems engineer to deal with this.

§3.9: Hierarchical thinking

§4.2: Nine-Window Diagram

§4.7: Architectures

Top-level functions have to be allocated to subsystems, creating interfaces between these subsystems. In this process, one can step to other hierarchical levels to investigate the functions that are needed to accomplish the top-level function.

Decomposition–composition thinking is on the one hand a formal, logic, process of determining what happens when a function is split: what interface(s) are created? This formal process and logic must be accompanied by thorough documentation, possibly supported by a computer tool that can check consistency and traceability[6]. This divide-and-conquer strategy is paramount in organizing the development process. On the other hand, considering the way of thinking of most designers and the management culture in organizations, the big picture gets easily lost. A less formal view by the systems engineer or system designer is then essential. He has to ask questions like:

§4.13: Documentation and reviewing

- How do we put this together?
- How do we check whether it will fit before shipping the parts and building on-site?
- Is there an easy way to see whether a module is finished?
- Can each module be tested before integrating into the system? If not, what test setup is needed?

Managing and monitoring the interfaces between parts, and between the electronics, mechanics, optics, and software are crucial in this matter.

The system designer has to be aware that the total system is not on most designers' minds. The total system should worry the system designer very much; in fact, it should be his main worry. The thinking tracks we present help too. Making mental and concrete pictures of the total system is essential. The tools mentioned in Chapter 4 can be helpful, The ones mentioned here in the margin are of particular use. Involvement of the designers in preparing these pictures, diagrams and models helps in avoiding problems. A simple way is to let the electronics designer draw the subsystem from his point of view, including the interfaces with mechanics and software. Then let the mechanical engineer do the same, and then the software engineer. In most cases there will

§4.9: FunKey architecting

§4.10: A3 Architecture overviews

§4.8: System Budgets

[6] Several requirement tracking tools are available in the market.

be discrepancies. If identified early, they are easily solved. It is important before to let them *draw*, do not leave it to describing a project in words.

Questions to ask the designers:

- If you start detailing the design, what information do you need from software, electronics, mechanics or optics?
- In what format do you expect that information?
- What kinds of connectors and fasteners do you use?
- If you have the first prototype, what hardware and software do you need to debug, test and validate it?

In particular, look for deadlock situations where developers have to wait for each others' information or need each others' deliverables.

3.9 HIERARCHICAL THINKING

The systems engineering process treated in §2.3 shows a hierarchical organization[7] from system to subsystem to assembly. This structure is also how many systems are organized. But this organization does not necessarily dictate the way the system's control functions. In hierarchical thinking, the system designer has to consider ranking authority, facilities and priorities of the system's parts.

§4.5.1: Function Trees

As an example, for a house heating system, several options can be designed:

1. Every room has a temperature sensor (or even several) and a room unit where the desired temperature can be set. Both the set temperature t_s and actual temperature t_a are sent to a central control unit where all values from all rooms are gathered, and the heater is controlled. A set of valves is controlled so that each room gets the heat that is needed; see Figure 3.6(a).
   This is known as *central control, central actuation, central generation.*
2. Using the room data above, a central heater is designed to provide constant heat flow. Each room unit directly controls a valve to control heat flow into the room; see Figure 3.6(b). This is known as *decentral control, decentral actuation and central generation.*
3. Using the room data above, a heater installed in each room is directly controlled by the room unit; see Figure 3.6(c). This is known as *decentral control, decentral actuation, and decentral generation.*

Of course, more variants can be devised, for instance when municipalities are generating the heat flow. As an exercise, add two or three variants and list their advantages and disadvantages including the variants mentioned above.

---

[7] Hierarchy: a system in which people or things are placed in a series of levels with different importance or status (Merriam Webster online, retrieved 20150615).

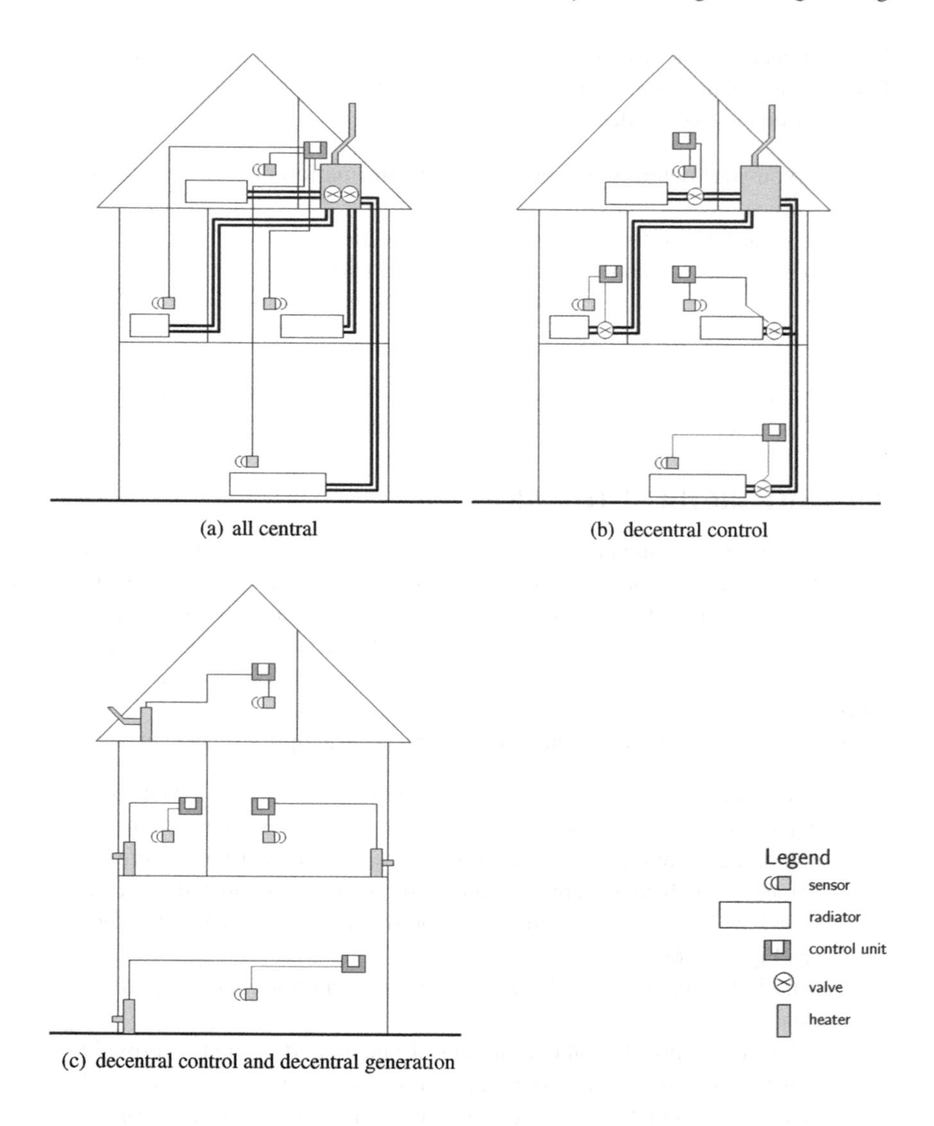

Figure 3.6 Three strategies for heating a house. In (a) generation, control and distribution are done centrally, in (b) generation is done centrally; but control and distribution are done decentrally. In (c) the generation is done decentrally with small heaters.

As seen in the example, the hierarchical reasoning can be applied to control, and also to generation and actuation. In a traditional car, there is only one engine. An electrical car can easily have an electrical motor in each wheel, thus providing four-wheel drive. Central control reduces the number of signals, but increases the number of power lines or power shafts. Localized control does the opposite. In computing there is the trend of cloud computing: storage of data and even programs *in the cloud* means storage on servers that can be accessed from anywhere via the Internet. The computer can be small and light (both in weight and processing power).

A final note on hierarchical thinking is that the system designer should be aware of hierarchies in organizations; both the company structure and the project structure. There *is* a line of responsibility similar to the line of command in an army. Dutch society is very egalitarian and most decisions are based on convincing and consensus. However, most cultures use a more formal way of decision making, with more use of hierarchy. In any case, a system designer often has no formal power but a good system designer will gain a lot of respect over the years. This will put him in a position of great influence and bring great responsibility. Combined with the fact that a system designer often works with uncertain information can yield a complex situation.

3.10 ORGANIZATIONAL THINKING

This book concentrates on the development of system (see Figure 2.1), but it should be noted that a system does not come to being without an organization that creates it. Most of what is said about the system under design is valid for the organization that creates the S.U.D. and therefore most thinking tracks above can—and should—be applied to the organization as well. There is an intricate relation between the project structure and the architecture of the system [Gulatti and Eppinger, 1996]. This relation may originally (if the system is not too complex) be directed from the system architecture to the organization. Once in place, the organization can influence or even dictate the system architecture. This phenomenon is tagged *Conway's law*, [Conway, 1968]. Figure 3.7 shows the ASML project structure in place while the first author was employed there.

Too many hierarchical layers in a project structure can hamper communication just as too many interface layers can hamper quick responses in a real-time control system. Depending on the size and required *agility* of the development, the organization structure should be relatively flat. It can be argued that a moderate level of chaos can result in very agile development teams, although this chaos requires a special kind of developers, and in particular system developers.

In any way, as a system is an interacting set of subsystems/components that use multiple disciplines, it is generally designed by multiple developers from multiple disciplines. The only way to achieve a well functioning system is to have them *communicate* intensively to create the necessary knowledge [Bonnema, 2014; Nonaka and Takeuchi, 1995]. The project organization and structure must facilitate this communication with frequent interaction, the creation of communication structures, development of *common ground* between the developers, and proper documentation of the knowledge generated.

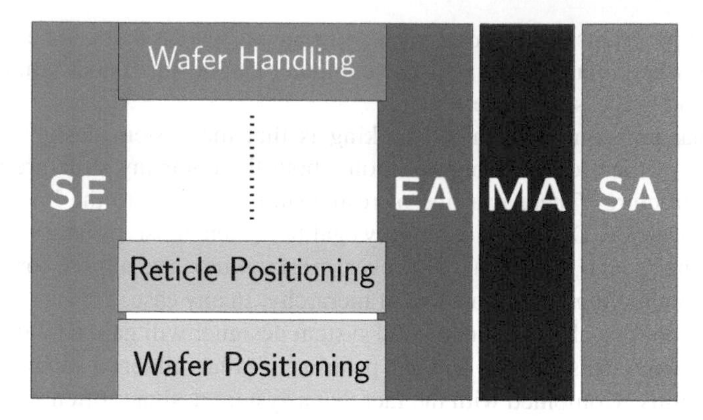

Figure 3.7 The ASML project structure. SE: systems engineering; EA: electrical architecture; MA: mechanical architecture; SA: software architecture. The horizontally shown projects deliver the main system functions (positioning and scanning the wafers or reticles, or handling them in and out the wafer scanner), while the vertically shown projects provide integration. SE covers the overall system integration; EA delivers the electrical infrastructure, MA the mechanical infrastructure and SA the software infrastructure.

3.11 LIFE-CYCLE THINKING

Life-cycle thinking can be understood in three distinctive, but equally important, ways:

1. The life-cycle idea → design → production → deployment → use → retirement
2. The life-cycle as in resource use and environmental impact
3. The life-cycle of the project that is put in place to develop, build and sustain the product

Each of these will be treated in separate subsections. The first is called the *product life-cycle*; the second is the *resource life-cycle*, the third is the *project life-cycle*.

3.11.1 PRODUCT LIFE-CYCLE THINKING

Every product goes through different phases in its life. Although names can differ, the following phases are recognized:

- Identification and/or idea generation
- Design
- Production
- Distribution, deployment, fielding
- Use
- Retirement

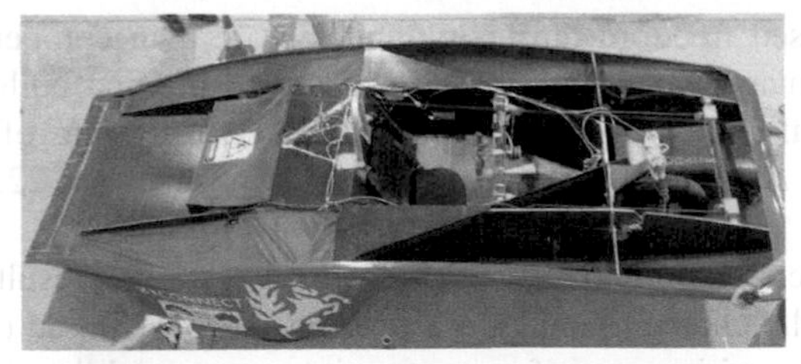

(a) The inside of the car, exposing the monocoque structure (detail of a photo by Gijs Versteeg)

(b) The mockup used to perform tests while the monocoque was being designed and manufactured

Figure 3.8 The 2011 Solarteam Twente car "21 Connect" using a monocoque structure. (Courtesy of Solarteam Twente)

For mass-production goods, emphasis has to be put on the production and distribution. For one-off products like space systems, distribution is less critical, but the way the system is put into use is critical. For civil engineering, the design and production phase are very critical; in particular the impact of the production phase has to be considered: the addition of an extra traffic lane for a highway must exert minimal impact on normal traffic during rush hour in the production phase.

Product life-cycle-thinking can be characterized by evaluating the impact of a design decision on all phases in the product life-cycle, much like standing in a *paternoster lift* that moves along different floors of a building. The floors resemble the phases in the product life-cycle, the person in the lift resembles the systems engineer evaluating a design decision.

In the design of the 2011 Twente solar car, it was decided to use a monocoque construction; see Figure 3.8(a) as this provides a light and stiff construction (advantages in the use phase). However, in the design and construction phases, consequences are that the monocoque frame cannot be designed before the aerodynamic shell design is finished. The monocoque frame is the first part needed in the construction phase as all other parts are connected to it. A solution was found in making a second frame of simple welded steel that would act as a test setup for most other parts; see Figure 3.8(b). Some additional work time was needed as most construction work had to be done twice.

The increased freedom during planning more than compensated for this small disadvantage; many lessons learned during the first build were utilized during the final build. The team followed several principles of the TRIZ tool set (Appendix A): 10, Prior action; 13, Do it in reverse, and 22, Convert harm into benefit.

The consideration of all product life-cycle phases often results in trade-offs, for instance, the choice of making a product more desirable (and thus more expensive) or requiring more frequent maintenance (and thus incurring lower purchase price). This decision may require negotiations with the customer.

3.11.2 RESOURCE LIFE-CYCLE THINKING

Producing products uses resources in the form of human involvement, energy use, material use, and time. At present, there is an increasing awareness of the impact this has on the sustainability of the way of the western life-style [MacKay, 2008, also an excellent example of modeling]. Taking into account the resource use, and whether it has irreversible consequences becomes essential in the system design process. Life-cycle analysis (LCA) is an often used tool. Cradle to cradle is an approach to ensure optimal use of resources.

The systems engineer, though not a specialist in this area, should be aware of the basic principles of LCA, as for instance outlined in [Baumann and Tillman 2004], and Cradle to cradle [Braungart and McDonough 2009], and think about resource use. Again, trade-offs will occur between using more resources in the development and production phase versus using more resources in the use phase.

3.11.3 ORGANIZATION LIFE-CYCLE THINKING

The organization also has a life-cycle just like the S.U.D. In order to create the S.U.D., an organization has to be designed, created, used, maintained and retired. Typically, a design organization starts small, with a team that investigates the need and creates a document stating the customer wish and the system specifications; see §2.3. This team is typically formed from the marketing and systems engineering staff, and supported by others if needed. The next phase in the organization life-cycle is, (see Figure 2.4), when the system design is created. This requires a larger team, and thus more facilities.

Even more organization is required in the next phase, when the split into subsystems is made. Then each subsystem gets its own staffing, and the number of people involved increases rapidly. The need for communication and documentation increases rapidly, and this needs to be anticipated by project management and the systems engineer(s) in the early phase.

§4.13: Documentation and reviewing

After the S.U.D. has been designed, it will be produced, distributed and put to use; it has to be maintained and has to be retired one day. This life-cycle has to be supported. Depending on the number of systems and the context of use, the team is small or large. For railway systems the maintenance team is very large, and the part of the total life-cycle cost spent on maintenance is significant. For instance in railway material, maintenance cost over the life span (30 years) is twice the purchase cost [van Dongen, 2011].

Also, the type of people involved changes. In the design phase, people with creative minds are needed, while in the maintenance phase, more practical people are needed. Note that this causes a problem, as information about situations that occur during the use phase and are dealt with by the maintenance-people may never reach the designers. Thus these situations may show up in the next generation in the same manner. Regular contact between "sustain" and "design" is therefore recommended.

In most civil engineering projects (the *design-bid-build* structure), the maintenance is often left to a totally different organization from those involved in the design and built phase(s). This complicates development and management as design choices that ease building may have a very negative impact on maintenance. Current developments in the direction of *design and construct* or *design-build-finance and maintenance* (or even *and operation*) type of contracts are expected to alleviate this.

From an organization life-cycle perspective, it is wise to mark the important milestones in the whole life-cycle. Having a party at times can help to improve cohesion in the project team, increase flow of information between different parts of the organization and increase motivation.

3.12 SAFETY THINKING

Product safety is paramount. There are many regulations to which products and systems must comply before they can be declared as safe. A prominent one is Directive 2006/42/EC on machinery[8]. During the development process, safety in all situations has to be ensured. With the constant pressure to save money, safety measures and procedures, or added components to ensure safety may not be given adequate attention. However, the consequences of ignoring or neglecting safety can be enormous. One recent example is the BP *Deepwater Horizon* disaster in 2010 [Wikipedia, 2015a]. Another shocking example is the series of events that led to the explosion of the *Piper Alpha* oil platform [Wikipedia, 2015b]. In the *Deepwater Horizon* case, procedures and regulations were neglected; in the *Piper Alpha* case, the procedures and regulations were followed too strictly and a common sense intervention would have prevented the largest part of the disaster.

§4.11: Failure mode and effect analysis

§4.12: Risk management

§4.4: Scenarios

The systems engineer has to reason about how the product can be used whether use is safe, in what ways the product can be used unsafely. Note that depending on the type of user (persons with specific training and responsibilities, or general public) the safety measures and the handling of exceptions

[8] http://eur-lex.europa.eu/legal-content/TXT/?uri=CELEX:02006L0042-20091215

can take a large amount of the development effort. Where normal behavior is the usual thinking model of an engineer, safety thinking requires that critical situations and so called *bad-weather behavior* has to be considered.

It may be wise to create points of failure, to avoid failure in awkward or dangerous parts of the system. Examples are circuit breakers in electrical systems. These are the weakest links in the circuit designed to prevent overheating or worse in other parts of the circuits. A mechanical example is a breaking pin (like a wooden pole) to connect a plow to a tractor. When the plow hits a rock, the pin breaks and the rest of the machinery remains intact.

3.13 RISK THINKING

Designing innovative products and systems comes with risks. Avoiding risks will automatically lead to dull and non-discerning products and thus to less profit or even loss. It is therefore a matter of managing the risks in the project such that their impact can be controlled and contained. Nevertheless, there is always a gambling component involved. Everyone involved in the development process must be aware of any risks that may develop, and the organization must have a mechanism in place to handle the risks identified.

Risks may occur in different ways, the most prominent ones for engineers being technical risks. These relate to uncertainties in reaching the required performance level, reliability and other technical requirements [Suh, 2005]. However, one needs to realize that there are different types of risks according to the INCOSE SEH Working Group [2008]:

§4.14: Modeling and simulation

Technical: the system or a part does not function or is not accurate, reliable, fast or efficient enough

Cost: development costs are not according to plan (this can be due to technical risks) or materials are more expensive than expected

Planning: the planning is not as expected: the delivery date is shifted forward or backward

Program: decisions are made outside the project, at management level; for instance the priority is shifted

Note that the risks are not isolated. A technical risk can lead to cost risks and vice versa. Other relations exist as well.

Handling these risks can be achieved in a variety of ways. Some of the most common ones are:

Avoiding: finding an alternative solution where the risk does not occur

Containing: analyzing the risk thoroughly, then taking each step very carefully, and making sure that everything is done to contain the risk from increasing and causing a chain of negative effects

Taking: taking the risk; this may be necessary when there is no alternative

Delegating: the risk is taken, but the consequences are made someone else's responsibility. This can be a good option when the development project is a large joint venture. Sometimes even the risk can be imposed upon the end user (read the end user-license agreement that accompanies software)

In all cases *time* is an important factor. The earlier a risk is identified, the more time there is to identify options and alternatives. Risk identification and analysis should be done as early as possible in the project. Modeling and simulation, experience and prototyping help in handling and reducing the potential impacts of risks.

§4.11: Failure mode and effect analysis

Risk thinking considers these different issues, reasons about the likelihood, impact and possible contingencies and mitigation scenarios.

§4.12: Risk management

3.14 SUMMARY

The thinking tracks presented above summarize years of experience and results from theory and scientific literature in different engineering disciplines. Some advice may look obvious or too general. Nevertheless, during a design project wandering along these thinking tracks will result in new insights and yield increased knowledge about the S.U.D. This, in turn will produce a better end result and/or avoid mistakes and/or reduce risks.

We find that experienced systems engineers continuously use the different thinking tracks (explicitly or not) to monitor, evaluate and steer the development process. For a novice systems engineer, this may be hard. A good strategy is then to switch from one thinking track to another at regular intervals. Over time, the intervals between the tracks will reduce. Also, one will find that some tracks are more relevant to a specific application or organization than others. It is good practice to review one's way of working regularly, so that the switching between and selection of thinking tracks can improve over time.

In the next chapter we present concrete tools that can help in using these tracks.

4 System Design Tools

In this chapter we introduce a number of practical tools that help support the thinking tracks and implement the systems engineering process.

4.1 INTRODUCTION

After discussing the systems engineering process in Chapter 2 and the thinking tracks in Chapter 3, it is now time to fill the systems engineer's toolbox. In this chapter we introduce tools that can be used at various places in the SE process, or to support the thinking tracks. Some tools are suitable throughout the process, others are more suited to solving problems as they occur. In the previous chapters reference was made to tools to be used at specific points in the process or the thinking tracks. Here introduction to the tools can be found. First, a few terms need to be introduced and/or defined.

Function: a specific or discrete action (or series of actions) necessary to achieve a given objective [Blanchard and Fabrycky, 2011, p. 100]. Functions and thinking in functions are important in systems engineering and systems design. The functionality of a system determines to a large extent the very reason for existence of the system the name of the product reflects this. A coffee maker makes coffee, a computer computes, a music player plays music. This also illustrates the way to describe a function: as a verb and a substantive. The verb describes the action and the substantive describes the object the action acts upon.

Stakeholder: anyone using the system or involved in the development process.

Utility or **infrastructure:** functions that provide service to other functions in the system without direct benefit to the customer.

4.2 NINE-WINDOW DIAGRAM

The nine-window diagram tool, Figure 4.1, originates from the theory of inventive problem solving known as TRIZ [Altshuller, 1997; Salamatov, 1999] (see Appendix A for a short general introduction on TRIZ). The tool is simple to use and puts the system to be developed in its temporal and hierarchical context. It gives the opportunity to describe the previous situation, the present situation and the envisioned future situation at the level of the system to be developed. By also looking at the lower hierarchical level, the implications on that level are shown. Even more interesting is to look at the higher hierarchical level (the supersystem-level or system of systems level). The nine-window

§3.9: Hierarchical thinking

§3.2: Dynamic thinking

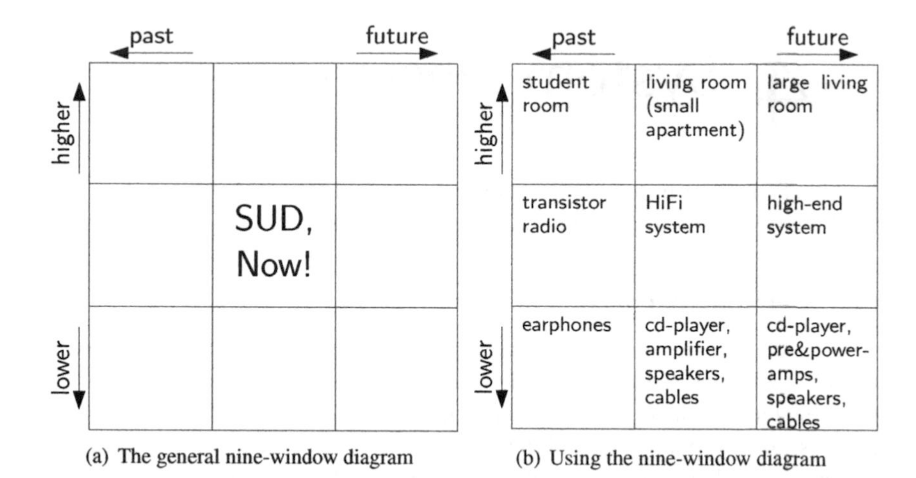

Figure 4.1 The nine-window diagram. Horizontal is the temporal development, vertical is the hierarchical development. In the center window is the S.U.D. as it is developed now.

diagram forces one to think about it such issues and promotes discussion of the consequences by the multidisciplinary team.

Another way of using the tool is supporting scenario-based design (see §4.4). Several versions for the middle (now) and right (future) columns are created. These can be discussed and evaluated based on criteria that resemble the customer wish. The best version is selected and elaborated into a system design.

The tool is well suited for discussions with diverse stakeholders like people from marketing, senior managers, system architects and specialists. Also note that variations can be made where one of the two axes is changed. Instead of the combination time and hierarchy, one can make a nine-window diagram of time and customer, or time and market. In Gadd [2011], more information on the use of the nine-window tool can be found.

Applied to the case of developing the solar racer, one version of the nine-window diagram can look like Figure 4.2. Here the 2007 and 2009 cases are depicted in the left and middle columns. The 2011 case is seen as a future case, as it was for the 2011 team when they started. Of particular interest is the fact that sponsor and supplier satisfaction is important in the supersystem row. Apparently it is in the next team's interest to keep these stakeholders satisfied before, during and after the race, to provide the next team with a good basis for a new development project.

4.3 CONTEXT DIAGRAM

The context diagram shows the system in its context. In its simplest form it looks like Figure 4.3 with the system under design in the center and the

	← past		future →
higher ↑	2007 regulations, sponsors & suppliers satisfaction	2009 regulations, sponsors & suppliers satisfaction	2011 regulations, technological developments, sponsors, suppliers, advisors
	2007 Solar Racer: "21", results	2009 Solar Racer "21Revolution", results	2011 Solar Racer
lower ↓	tilt-panel, Fresnel lenses, 2 front wheels, one driven rear wheel	tilt-panel, Fresnel lenses, 1 driven front wheel	rigid body, 6m^2 silicon cells, optimized cell pattern

Figure 4.2 The nine-window diagram of the solar race car team depicting past and future features.

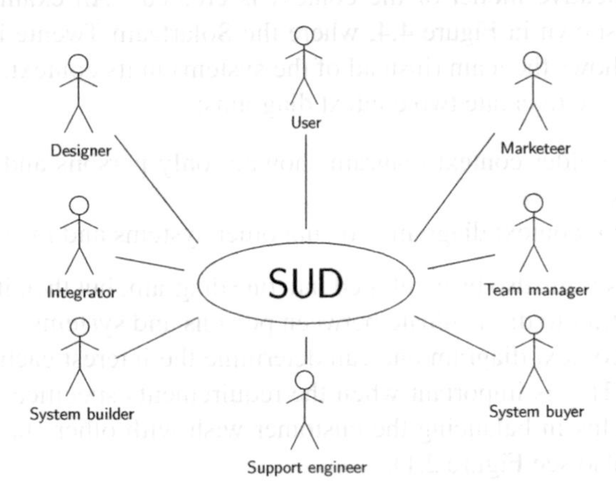

Figure 4.3 Context diagram template showing the stakeholders around a system under design (S.U.D.). Each of the stakeholders must be detailed. Is the user a skilled technician, or member of the general public such as a non-professional car driver. In an extended context diagram, other systems can be shown as well.

stakeholders around it. The stakeholders then have to be named and detailed with who they are and what they require of the system. Stakeholders can be organizations or persons. In case of consumer products, the context diagram shows persons like the end user and non-users, and their interests are mostly represented by a marketing department. For a business-to-business system like a wafer stepper or a fighter jet, the context diagram will show organizations like the procurement office. Then it is easier to talk to the end users as they are fewer.

§3.5: Operational thinking

§3.9: Hierarchical thinking

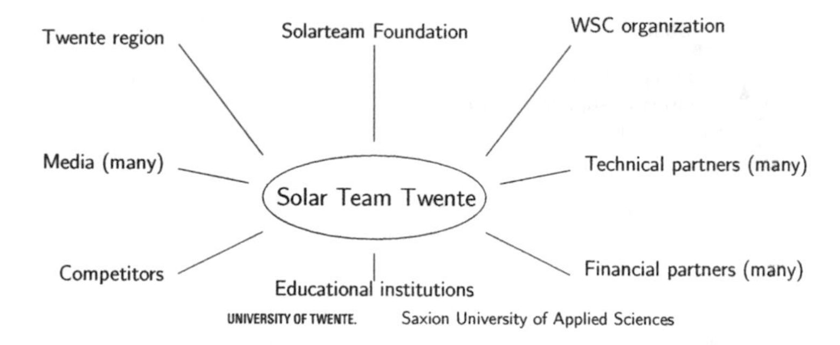

Figure 4.4 Context diagram for Solarteam Twente.

The context diagram can be extended by showing other systems around the S.U.D. When simple pictures are taken to represent these other systems, a rich and communicative model of the context is created. An example at a very high level is shown in Figure 4.4, where the Solarteam Twente is taken as an example. It shows the team (instead of the system) in its context.

It can be wise to create two context diagrams:

1. A stakeholder context diagram showing only persons and groups of persons
2. A system context diagram showing other systems and infrastructures

Of course these can also be combined into one diagram, but then it may be wise to use formatting to discriminate between persons and systems.

From the context diagram one can determine the interest each of the stakeholders has. This is important when the requirements specification is not yet defined. It helps in balancing the customer wish with other stakeholders' requirements (also see Figure 2.1).

4.4 SCENARIOS

In Muller 2004a; Wojtkowski and Wojtkowski 2002, the usage of *story telling* in systems engineering is treated. In other places, the technique is called the use of *scenarios*. A story or scenario is a textual description of the use of the system to be designed. As an example the following story explains the morning preparations of the solar team while camping in the dessert.

> Half an hour before the sun rises, the *panel crew* wakes and checks the GPS and compass data in order to know exactly where the sun will rise above the horizon. They move back and forth to make sure they find the lowest spot on the horizon, and arrange the panel stand, battery and electronics, ready to catch the first rays of sunlight.
>
> A little later the rest of the crew wake up to start the morning rituals: make breakfast, break down the tents for the facility team; fire up the com-

puters and get the latest weather data via satellite for the team manager and strategy planner. The technicians start the car electronically and do the functional and safety checks.

The team breakfast at 6:30 is combined with the day briefing. At 7:00 the breakfast is over and the weather car leaves the site. The campsite is cleared, the first solar car driver prepares herself and the convoy is assembled and is ready to start at 7:45, when the final safety check is done. However, because of regulations the convoy cannot start before 8:00. At 8:00 the solar car and the two accompanying cars (the escort car directly before, and the race control car directly behind) drive off. The speed is controlled from the race control car, based on the weather information received earlier and sent.

The rest of the team packs the last items into the truck and the other cars also prepare for departure, but their schedule is less critical.

While overtaking a road train, it starts to accelerate. The driver of the solar car notices this. She overrides the cruise control and increases speed somewhat, knowing it will cost extra energy, but safety is more important than any other factor. The overtaking maneuver is completed successfully, and the speed control override is ended; the solar car is back to normal cruise control operation. The strategy model is updated to take the extra energy used while overtaking into account. After a few seconds of number crunching, the cruise speed setpoint is adjusted.

This story has been written based on the solar team and context knowledge. The form of a story provides a rich and easy to interpret description that makes discussions among colleagues easy. Before a story is analyzed it should have been discussed among the designers, and whenever possible with the end users and other stakeholders. Discussions with the latter groups are possible because the use of jargon is kept to a minimum in such scenarios.

Based on this story, a state transition diagram like Figure 4.7 can be drafted. There is a close relation with operational thinking; in fact the scenario is the result of and input to operational thinking.

§3.5: Operational thinking

Another story about the solar car describes its entire life-cycle, from its design to its retirement. In that case the functions that relate to the production, distribution, use and retirement can be found. For a coffee cup this is relatively simple. For complex systems like cars, aircraft or wafer steppers, this can be a complicated story. In particular production will pose many problems that can be found early in the design process by creating and analyzing these stories. Thus for each product several stories can and should be written.

From a story as given above, software engineers tend to derive *use cases*. These are short descriptions of the usage of the software system. In Daniels and Bahill 2004, the application of use cases as a complement to the "shall" requirements common in systems engineering is treated.

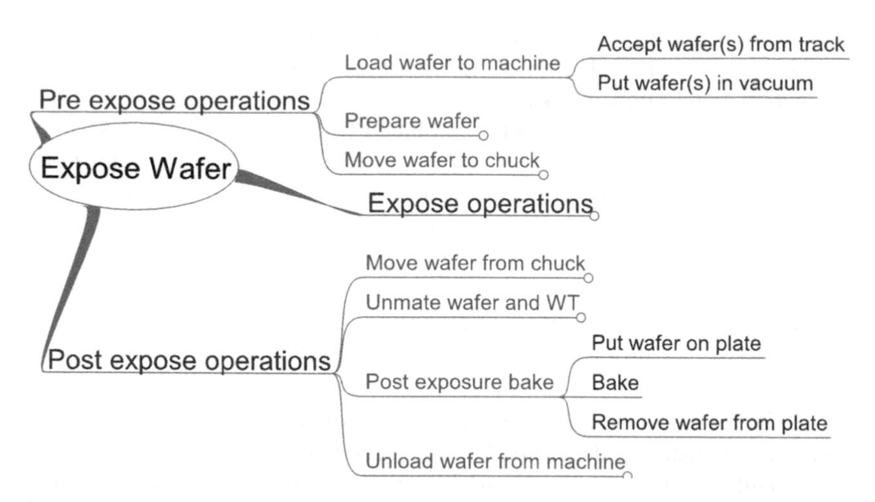

Figure 4.5 An example of a function tree that serves as an inventory of functions for a wafer stepper [Bonnema, 2011].

4.5 FUNCTIONAL MODELING AND ANALYSIS

Functions play an important role in systems design and engineering. Therefore modeling the functions from the earliest possible stage in the project is essential. Many different formalisms have been developed. In this book we treat only a few that are most common, or are very useful in a multidisciplinary context:

- Function trees and mindmaps
- Influence diagrams
- Functional block diagrams (FBDs)
- State transition diagrams (STDs)

4.5.1 FUNCTION TREES

When starting to design a new system or planning an upgrade to an existing system, it is useful to create an inventory of what the system has to do. Starting from the highest level, one can make a tree of functions like the one shown in Figure 4.5. Mindmapping software can be very useful for this. There are commercial programs, but the Freemind open source package (`http://freemind.sourceforge.net/`) is good and can be used on several platforms.

§3.9: Hierarchical thinking

By making a mindmap, one can subsequently create hierarchical levels that can later be used to define subsystems. With a tool like Freemind, one can hide lower branches so that discussions can focus on a specific part of the tree. Also, complete branches can be cut into a new file that can be used for further, lower level elaboration. This way, hierarchical thinking is supported easily.

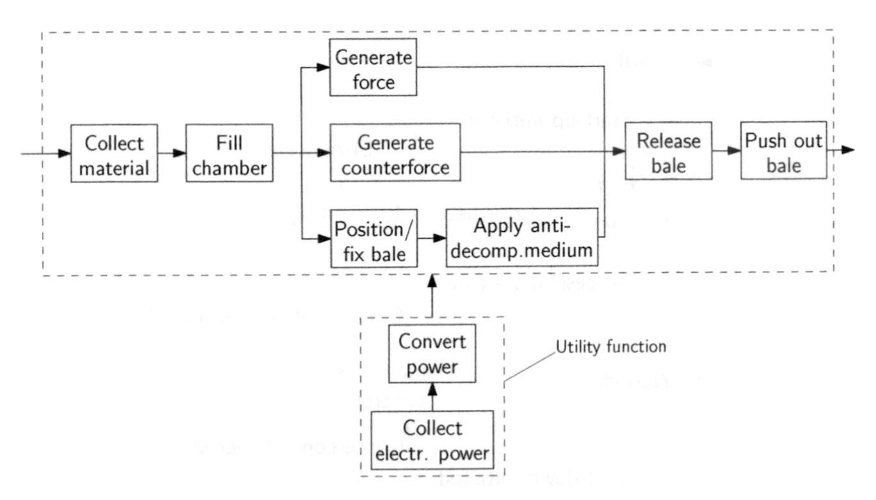

Figure 4.6 An example of a functional block diagram showing the functions of a waste baler. The system is used to compress waste [Bonnema, 2011].

4.5.2 FUNCTIONAL BLOCK DIAGRAMS

The functional block diagram (FBD) is possibly the model that is most widely used by the systems engineer, together with a requirement specification. It is a series of blocks that represent the functions the system has to perform sequentially or in parallel. The FBD can focus on any hierarchical or life-cycle level, and even an entire project. So there are FBDs on life-cycle functions, showing a satellite's life from development to launch, mission to re-entry and landing, and FBDs that span only seconds or micro-seconds when they describe a series of events that form a computer interface or protocol.

§3.11: Life-cycle thinking

§3.9: Hierarchical thinking

An example of a simple FBD is shown in Figure 4.6. This shows that functions can be sequential and parallel and utility functions can be modeled. These do not contribute directly to the output of the system. However, they are needed to ensure proper operation of the other functions. Figure 3.5 is also a functional block diagram illustrating with feedback.

§3.3: Feedback thinking

The FBD is most suited when there is one main track and the process depicted runs from left to right. There are possibilities to show parallel functions (see Figure 4.6), and even choices (by incorporating "OR" items). Also, hierarchy can be shown, as it is possible to work out one or more functions in one or more separate diagrams. The FBD is limited though, when in the sequence of the functions is unknown. This occurs when there is user interaction or non-predictable interaction with other systems is unknown. In that case the state transition diagram is better suited.

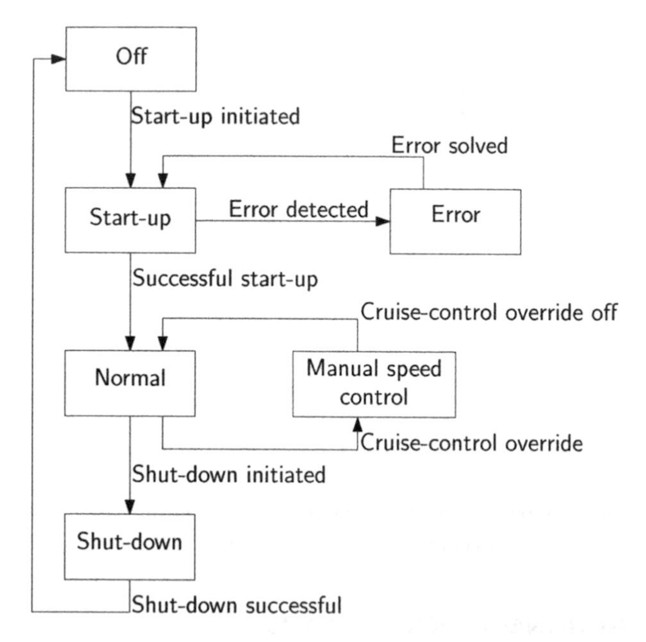

Figure 4.7 A state transition diagram for the operation of a solar racer. It shows the top level operation of the system, including start-up and shut-down. Every block (except Off) has to be worked out further in a new state transition diagram.

4.5.3 STATE TRANSITION DIAGRAMS

The state transition diagram (STD) is a tool to investigate and design the behavior of a system. For this, states and transitions are defined:

state: a mode of operation or being
transition: passing from one state to another state

§3.5: Operational thinking

It should be noted that in a state several functions can be performed in parallel or sequentially. States may be analyzed further (using other STDs) and found to consist of several other states.

In a STD, the states are modeled as boxes, and the transitions are shown as arrows between the boxes. For every transition, the condition is given when it occurs. Figure 4.7 shows a basic STD for a solar racer. It shows the outcome of discussions determining that there is a cruise control and that it can be overridden. A next step can be to further elaborate on the "start-up", "normal" and "shut-down" states. Also, the error handling has to be worked out further.

Note that every state needs to have at least one transition coming in, and one going out.[1] The STD can reveal possible *deadlock* or one-way situations. It is

[1]Notable exceptions may be weapon systems that stop existing after they have reached their goals.

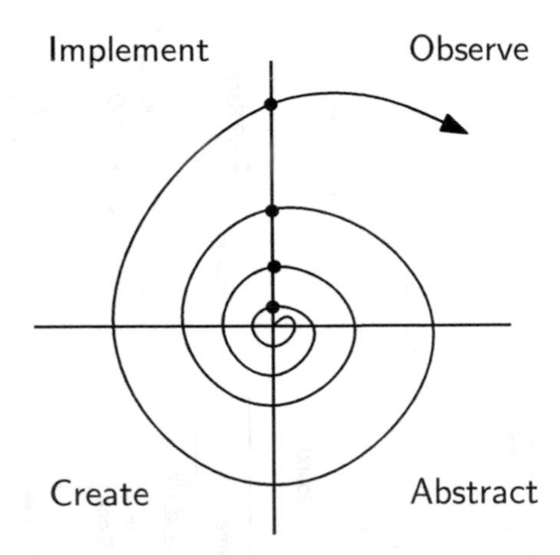

Figure 4.8 Spiral development model.

easy to make a simulation based on such a STD, to check whether the behavior is logical and understandable for the target user.

The use of STDs where parts are implemented in prototype software (for instance for investigating user interaction), supports a development approach called "spiral development", see Figure 4.8.

4.5.4 INFLUENCE DIAGRAMS

There are two types of influence diagrams: a pictorial investigation of influences and a more logical approach.

A pictorial influence diagram is shown in Figure 4.9. It depicts a system under consideration (like an iconic diagram, a sketch, block diagram) with all effects influencing the system indicated with text and/or symbols. The influences are indicated with either ellipses or clouds. An ellipse is used when the place of influence can be clearly indicated; a cloud is used in other cases.

Figure 4.9 shows a wafer stepper or scanner. Here *overlay* is one of the key performance indicators. It describes the accuracy between the lateral position of two layers of an integrated circuit. Accurate overlay can only be achieved by carefully selecting working principles and checking every decision against its possible influence on overlay. Note that overlay does not map onto one or a few specific functions or parts of the wafer stepper; it is a characteristic of the system as a whole. In Figure 4.9 an inventory of the mechanisms that contribute to overlay are shown (also see Muller 2004b). The diagram was made by the first author to create an *overlay-budget* (§4.8) on the one hand, and to raise awareness among the designers about the importance of designing with overlay in mind on the other hand.

§3.6: Scales thinking

§3.7: Scientific thinking

The logical influence diagram models the way in which variables and effects relate to each other. One can say that such an influence diagram is a visualiza-

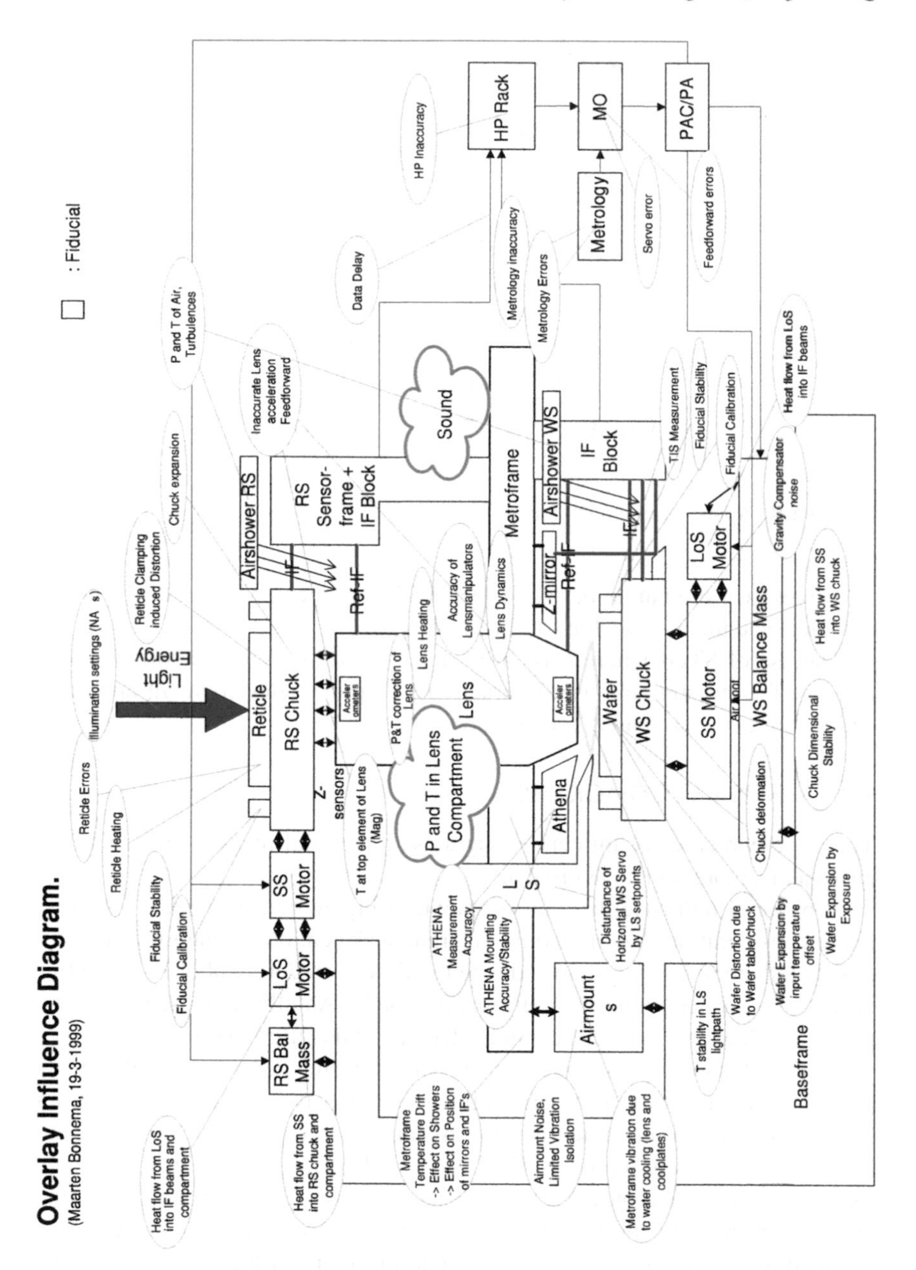

Figure 4.9 Concept of a wafer scanner: items influencing overlay are shown as clouds and ellipses.

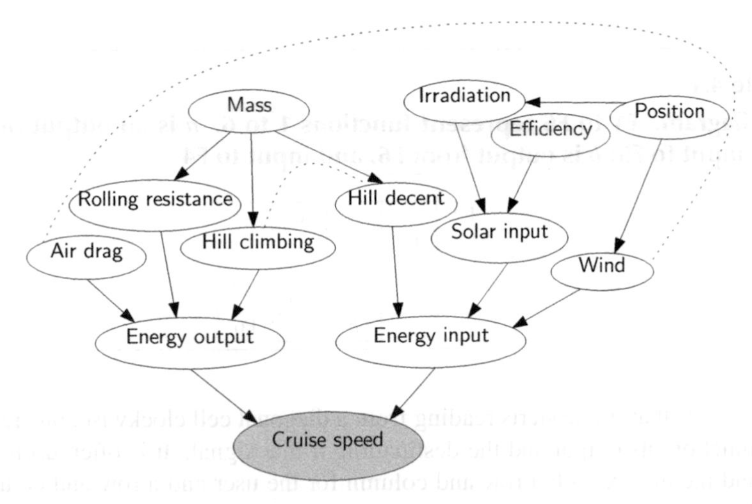

Figure 4.10 Logical influence diagram for a solar racer. It models the contributors to the cruise speed, *i.e.*, the speed at which input and output of energy balance. The dotted lines indicate linked items.

tion of (a chain of) formulas. An example is shown in Figure 4.10 where the effects that influence the cruise speed of a solar racer are investigated.

§3.9: Hierarchical thinking

Such a diagram is created starting from the effect to be investigated (shown at the bottom). Then, the model is improved by subsequently asking what variables and parameters do I need to calculate this effect? In Blessing and Chakrabarti [2009] the reference model is used to do almost the same. At the arrows, references to sources of information are made, so that different types of effects—even contradicting effects—can be modeled. Also, fault trees look similar. They can be used to investigate risks and errors.

4.6 N^2 DIAGRAM

The N^2 diagram (also see Kapurch and Rainwater [2007]) is a useful tool in different stages of the architecting process with respect to interface management:

§3.8: Decomposition–composition thinking

- To inventory all interfaces between functions
- To compare architectures regarding the number and type of interfaces
- To monitor the interfaces when the system is developed further

By its nature the N^2 diagram aids in working through all possible interfaces, minimizing the chance of omitting one. The N^2 diagram consists of:

- A square matrix (thus its name) with the functions on the diagonal
- The *outputs* of a function listed on the row of that function
- The *inputs* to a function listed in the column of that function

Table 4.1

N^2 diagram. F1 to F6 represent functions 1 to 6. a is an output of F2 and input to F5; b is output from F6, and input to F4

F1					
	F2			a	
		F3			
			F4		
				F5	
			b		F6

This means that if one starts reading from a diagonal cell clockwise, one reads the function, its output and the destination of that signal. It is often useful to expand the matrix with a row and column for the user and a row and column for the environment.

This *functional* N^2 diagram is useful for analyzing the required interfaces. Once this is done, the functions will be allocated to subsystems. Then another N^2 diagram is made: the *modular* N^2 diagram. Here, the subsystems are put on the diagonal, and the interfaces between the subsystems are represented off-diagonal; see Table 4.2. A heuristic [Maier and Rechtin, 2000, p. 273–274] states that a sensible architecture has low complexity *between* the subsystems, and high complexity *inside* the subsystems. This is seen in the modular N^2 diagram as only a few filled off-diagonal cells.

§3.9: Hierarchical thinking

Note that N^2 diagrams can be made hierarchically so that one top-level diagram represents the system interfaces. Then, for each subsystem the interfaces inside this subsystem are worked out in a separate N^2 diagram. This way of working should consider the earlier mentioned systems architecting heuristic of low inter-module complexity, high intra-module complexity.

4.7 ARCHITECTURES

In creating large and/or complex systems the term *architecture* is often used. This term originates from building and can have many meanings. Defining it is not an easy task [Maier and Rechtin, 2000]. For this text, we will use the following definition which is based upon [Gulatti and Eppinger, 1996]:

§3.8: Decomposition–composition thinking

System architecture defines the parts constituting a system and allocates the system's functions and performance over its parts, its user, its super system and the environment in order to meet system requirements.

In other words, the architecture defines how the S.U.D. is split into smaller pieces to ease the development, but at the same time, it also defines how the S.U.D. is positioned in the environment, and how it interacts with the user(s), as already shown in Figure 2.1.

Table 4.2
N^2 diagram for a teletext to xml converter used to fill the electronic program guide (EPG) of a personal video recorder. The cells on the diagonal show the main components. Note that there are no interfaces below the diagonal

Environment	antenna signal; EPGinput.txt; Filters (NL123.vtf, ARD.vtf, ...)	channelIDs.txt	
	Receive teletext and filter unwanted info	EPG data.txt	
		Convert time, program name and add station info	EPG.xml
			PVR app

Note that any system of some complexity *has* an architecture, but it may not be created consciously or documented explicitly. It may, for instance, be in the head of only one person. It is in general needed to *document* the architecture. This requires several representations in different formats. Nearly all of the tools in this chapter are useful as representations for one or more aspects of the architecture. The combination of all different representations form the architecture description. The relevant standards IEEE1471 and the superseding ISO/IEC/IEEE 42010 make this discrimination between architecture, architectural description, views and viewpoints. (Note that it is not always required to use all representations; a wise selection has to be made.)

§3.10: Organizational thinking

The architecture has to address the following issues:

- How is the SUD divided into smaller, more manageable parts called *subsystems*?
- What are the *interfaces* between the subsystems?
- How are the functions allocated over the (sub)systems?
- How is the S.U.D. positioned in the *environment* and the rest of the world?
- What are the interfaces with the environment?
- How are the development paths between the subsystems (de-)coupled?
- Can parts be reused (for instance from previously developed systems or from systems developed by others)?
- What is implemented in hardware and software?
- What are the long-lead items (see Page 78)?

With careful architecting, for instance as described in §4.9, technical and business opportunities can be identified. For a given function, one can ask

§3.6: Scales thinking

whether opportunities should be explored by the system, the user, the environment, or a service provider and then determine whether profit can be generated from that. As an example, selling roasted coffee beans will result in less margin than selling portions of sealed, ground espresso coffee. This is also the razor-blade marketing philosophy. Here the razor holders are cheap or even free. Profit is made on the blades. Although they are not very expensive, they are sold in large quantities over long periods of time. Thus a modest margin on each piece adds up to large sums for the company selling them.

4.8 SYSTEM BUDGETS

The word *budget* makes most people think of money. A budget is "a quantity (as of energy or water) involved in, available for, or assignable to a particular situation".[2] In finance, a budget also shows the breakdown of the diverse expenditures. The same holds for a system budget: it states the available quantity, and the way it is divided over the expenditures. In systems design, it can be used for items other than money. System budgets are used to allocate portions of the system requirements to subsystems, assemblies, components and parts. Requirements that can be divided using budgets are:

- Mass
- Power consumption
- Space
- Waste production
- Production time
- Service time
- Accuracy
- Productivity

In fact, most system requirements can be and should be divided over the subsystems using budgets. Some requirements like conformity to standards and safety regulations cannot be measured this way.

§3.6: Scales thinking

§3.8: Decomposition–composition thinking

§3.4: Specific-generic thinking

The budget should be prepared early in the development process, and should be the development goal for each subsystem. The basis for the budget is a combination of:

- Modeling
- Experience with previous systems
- Experiments
- Extrapolation of past results
- *guesstimation*[3]

Based on these activities, a first version that may only be one or two levels deep is devised by the systems engineer. Through interviews with experts and stakeholders, refinements are made. Possible conflicts can be identified and should

[2]Merriam-Webster online, retrieved 20150616.

[3]A guesstimate may be a first rough approximation pending a more accurate estimate, or it may be an educated guess at a solution for which no better information will become available. Source: http://en.wikipedia.org/wiki/Guesstimate

Table 4.3
A mass budget for a solar racer. Note that the structure is real, but the numbers are modified for confidential reasons

Item	mass [kg]	remark
EL powertrain	32	
EL telemetry	3	
EL solarpanel	8	
monocoque	65	
suspension	18	
wheels	30	
steering	4	
braking	4	
detailing	9	
driver	82	80kg by regulations, plus harness and cushion
margin	5	managed by SE
Total	260	including driver
	180	excluding driver

be solved through further modeling and analysis and/or discussions with the corresponding stakeholders. The result is a budget that is accepted by all stakeholders. It can be useful to represent the data in a tree-form that corresponds with the system breakdown into subsystems.

As shown in Figure 3.2, in the early phase when the first budgets are made, there is some uncertainty. When more information becomes available during the development process, adjustments to the budget may be necessary. There can be transfers from one item to another or adjustments to the total budget. Therefore, a margin should be included in the budget. This can be done implicitly or explicitly:

Explicit margin: An example is shown in Table 4.3. A margin is included for the total budget item. Every subsystem must adhere to its budget. The margin can be applied in cases of overruns. The system engineer manages the margin.

Implicit margin: No margins are included in a budget. The systems engineer should know where additional room can be found or transferred. If a subsystem encounters budget problems, the total budget may have to be increased.

Both solutions have advantages and disadvantages that have mostly to do with the culture in the organization. An implicit margin appears to be more strict. However, once the flexibility is known, the assumed strictness is gone.

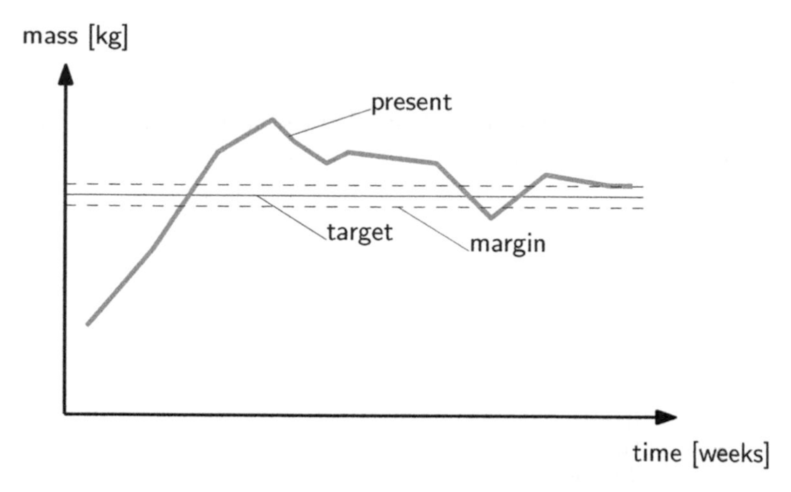

Figure 4.11 Tracking a mass budget. The thin black line shows the target value. The thick grey line shows the present value at different moments in time based on the design at these moments. At the beginning, missing data leads to a very low mass. Later the target is exceeded.

Managing the budget during the design process can be done in a graphical manner as shown in Figure 4.11. Here the value based on the current design knowledge is plotted against time in the development process in a graph that also shows the design target.

4.9 FUNKEY ARCHITECTING

§3.4: Specific-generic thinking

§3.6: Scales thinking

In FunKey Architecting [Bonnema, 2008, 2011], a combination is made between the functions a system has to perform (§4.5) and the performance it has to achieve (§4.8) to balance the budgets. At the same time by making explicit links between the functions and the stakeholder's benefit through the use of key drivers, the contribution of every engineer is made explicit.

In addition to the earlier defined functions and budgets, we thus need to define the term "key driver":

Key driver: a generalized requirement representing a stakeholder's main interest

Examples of key drivers are *total race time* and *media exposure* for a solar racer, *cost per passenger per mile* and *turn-around time* for a passenger aircraft. For a system, the goal should be to define five to ten key drivers that should be quantifiable. The set of key drivers should span the set of stakeholders. On the one hand, the key drivers show the desired (required) performance. On the other hand, they express the engineering achievements. That way, the key

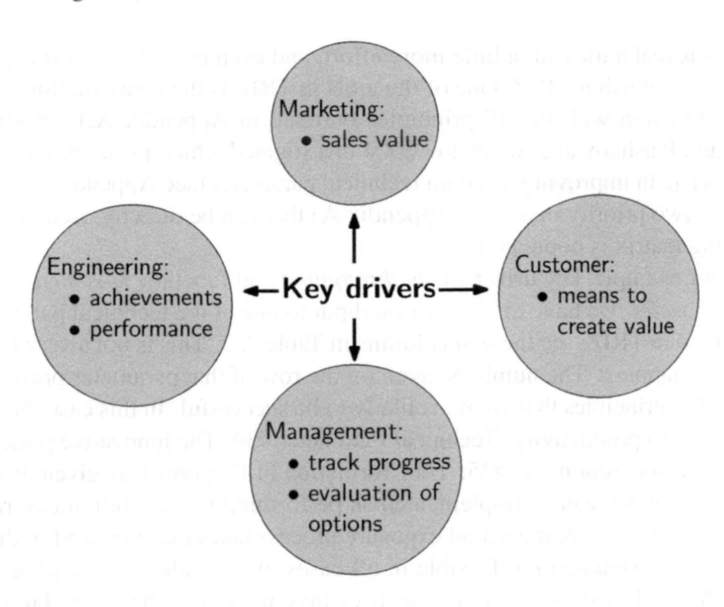

Figure 4.12 Key drivers act as interfaces between stakeholders.

drivers act as interfaces between the customer and builder of the S.U.D.; see Figure 4.12.

Inventorying the key drivers has to be done with care. The function of key drivers as interfaces between various stakeholders requires interviews and discussions with these various stakeholders. In a number of sessions, the systems engineer can update and finally fix the key drivers to be used for the S.U.D.

Next step is to determine which system function contributes to which key driver. For this, a *coupling matrix* $\mathscr{C}$ is used as shown in Table 4.4. A cross (×) indicates whether the function contributes as positively or negatively to the key driver. At first, this is a mere qualitative inventory. However, in a later stage, a quantitative analysis can follow. But before that, the qualitative analysis can

§3.8: Decomposition–composition thinking

Table 4.4
FunKey coupling matrix $\mathscr{C}$ used to investigate contributions of functions to key drivers (kd_i) [Bonnema, 2011]

	kd_1	kd_2	...	kd_m
function f_1			×	×
function f_2	×	×		
...		×		
function f_n	×	×		×

already reveal a lot with a little more effort, and even provide clues for system improvement using TRIZ. One of the tools in TRIZ is the contradiction matrix in combination with the 40 principles outlined in Appendix A.1. Bonnema 2008 and Ivashkov and Souchkov 2004 investigated which principles are most appropriate in improving a certain technical parameter (see Appendix 1.1). The result is two priority matrices (Appendix A) that can be directly used once the coupling matrix is populated.

If, for example, key driver kd_2 is *throughput* and function f_2 is *measure position of wafer*, we have to relate throughput to one of the technical parameters as defined in TRIZ; see the first columns in Table A.3. This is not a very formal or strict mapping. The numbers given on the row of that parameter present the innovative principles that are most likely to be successful. In this case throughput relates to productivity: Technical Parameter #39. The innovative principles are #10: Prior Action and #35: Transformation of Properties, as given in Table A.2. Prior action can be implemented as performing the function measure position of wafer before the actual exposure process takes place instead of during production. Although not feasible in all cases, it is a valid way of improving throughput. Transformation of properties may work in some cases, but is not seen feasible in this example.

4.9.1 COUPLING MATRIX TO BUDGETS

The coupling matrix $\mathscr{C}$ is a good basis for further refinement of budgets. The crosses in Table 4.4 can be further analyzed and replaced by numbers. These numbers can be percentages of the total available mass, time, money and memory needed to balance the total design. Another approach is to fill in known values from existing systems or the present state of the technology and compare the total with the performance to be achieved. Any tension is then made visible. Also, the field where the improvement should be found is clearly shown.

4.9.2 FUNKEY AS TRACKING TOOL

Like the use of budgets shown in Figure 4.11, the coupling matrix $\mathscr{C}$, once filled with numbers, can be used for tracking the design process. By doubling each key driver column into *aim* and *actual* columns, the development is tracked based on numbers. In Bonnema 2008 the use of Symmetrical Triangular Fuzzy Numbers (STFNs) is proposed to make uncertainty visible.

For more in depth information on FunKey architecting and its application, the reader is referred to Bonnema 2008, 2011.

4.10 A3 ARCHITECTURE OVERVIEWS

Systems engineers produce architectures. These are mostly represented in diagrams and documents. Standards for architectures like SysML are available [Object Management Group]. Yet, the most used tools by architects are word processors, spreadsheet programs and presentation programs. The diagrams and tables are put together into an architecture document with text. These documents are not often consulted after they are produced, reviewed and accepted because asking the architect for the required information is faster than looking it up in the documents.

A recent development [Borches and Bonnema, 2010; Borches Juzgado, 2010; Wiulsrød et al., 2011] is to present architectures in a compact and visual format so that the architecture information is readily available. The philosophy is to make the representation as simple as possible, but not simpler.[4] This is achieved by strictly limiting the available space to two sides of an A3 sheet of paper.[5] One side shows three types of models: functional, physical and quantification (Figure 4.13(b)), the other shows a textual summary and explanation (Figure 4.13(a)).

The title of the A3 architecture overview (A3AO for short) shown on both sides is very important as it describes, defines and limits the scope of the document. It should contain the S.U.D. and the specific aspect that is modelled and described. If this is not done, it will become impossible to stay within the bounds of one A3 sheet.

§3.6: Scales thinking

§3.8: Decomposition–composition thinking

§3.9: Hierarchical thinking

The model side of the A3AO comprises three main views:

- A functional view, showing the functionality of the system of interest (shown at the left in Figure 4.13(b))
- A physical view, showing how the functionality is divided over the system's components, and the physical components[6] (shown at the bottom right in Figure 4.13(b))
- A quantification view that shows the main performance numbers, similar to the system budgets (§4.8) and FunKey (§4.9) (shown at the top right in Figure 4.13(b))

Additional models can be used to support these views. These are shown in the middle section of Figure 4.13(b).

The textual side (Figure 4.13(a)) shows, each in its own box (from top to bottom):

[4] Paraphrasing Albert Einstein.

[5] A3 is an international paper size standard. 297 × 420 (American equivalent of 11 × 17 inches).

[6] The physical view for a software system shows the way in which modules and components are defined.

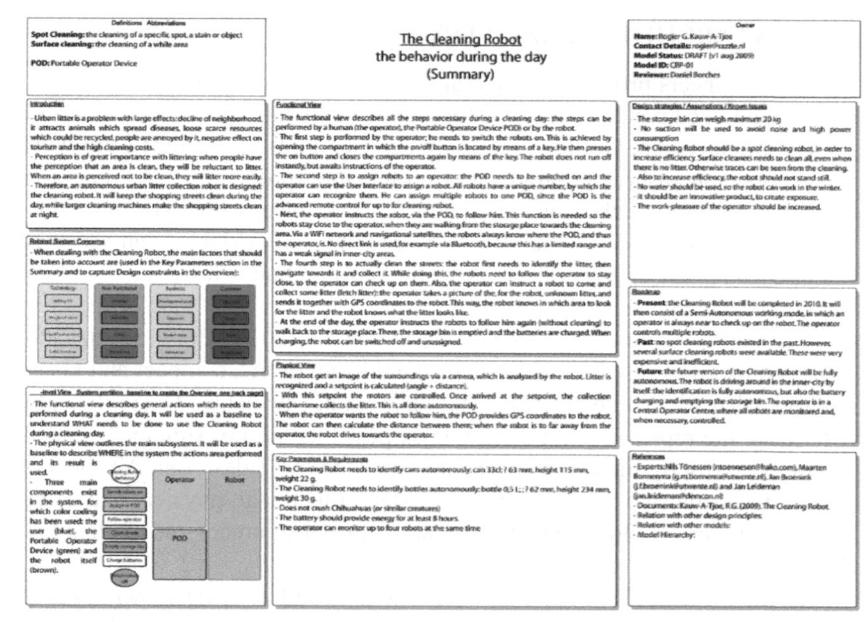

(a) Front side showing summary text side of the A3.

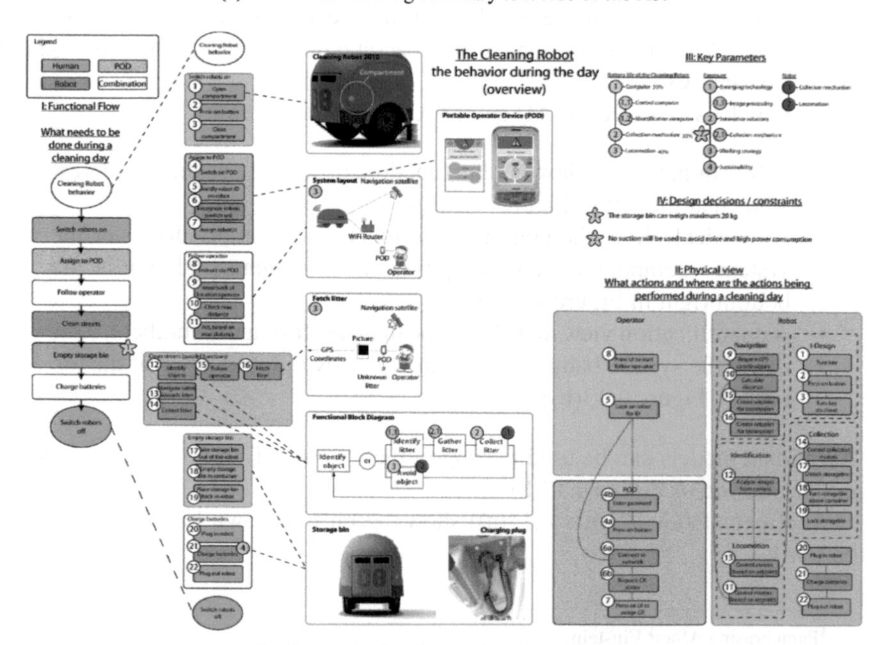

(b) Rear side showing model side of the A3

Figure 4.13 A3 Architecture overview of a litter-collecting robot [Kauw-A-Tjoe, 2009].

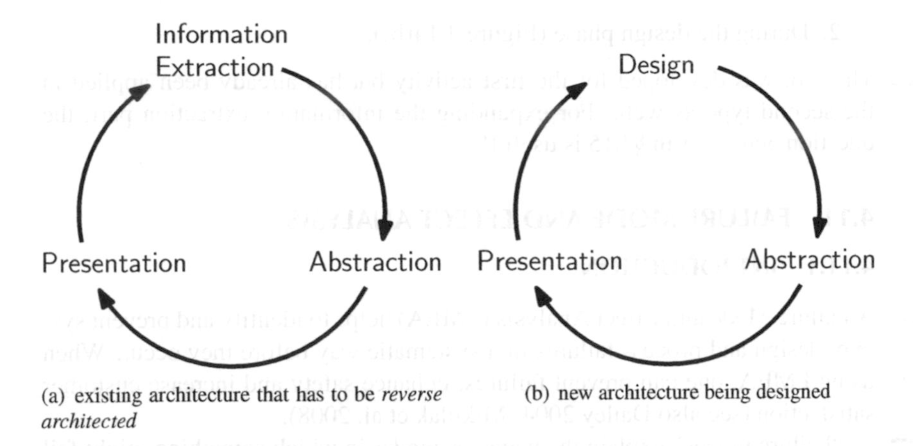

(a) existing architecture that has to be *reverse architected* (b) new architecture being designed

Figure 4.14 The A3 Architecture overview creation process (adapted from Borches Juzgado 2010).

Left column:

- Definitions and abbreviations.
- An introduction describing this A3O.
- The related system concerns that should be taken into account when creating the architecture. These follow from several of the system thinking tracks presented in Chapter 3. The system concerns generally are used for several A3AOs of the same (type of) S.U.D.
- The top level view showing the main partitioning. In this case how the functionality of cleaning litter is divided over the robot, the operator and the remote controller.

Middle column:

- Additional information for the functional view on the model side.
- Additional information for the physical view
- Information for parameters and requirements relating to the quantification view.

Right column:

- Logistics information: document owner, creation date, contributors etc.
- Design strategies, assumptions and known issues.
- Roadmap for further development.
- References to more information and experts.

Creating an A3AO is an intensive cyclic process that normally spans several cycles [Borches Juzgado, 2010]. It can occur in two different situations:

1. Reverse architecting: the architecture is already created, but the documentation is lost or not readily accessible (Figure 4.14(a)).

2. During the design phase (Figure 4.14(b)).

§4.15: Question generator

The tool was developed for the first activity but has already been applied in the second type as well. For expanding the information extraction part, the question generator in §4.15 is useful!

4.11 FAILURE MODE AND EFFECT ANALYSIS

4.11.1 INTRODUCTION

§3.10: Organizational thinking

§3.13: Risk thinking

A Failure Mode and Effect Analysis (FMEA) helps to identify and prevent system, design and process failures in a systematic way before they occur. When using FMEA, one can prevent failures, enhance safety and increase customer satisfaction (see also Dailey 2004; Mikulak et al. 2008).

"Failure modes" explain the ways, or *modes* in which something might fail in fulfilling functions and failures that affect the customer satisfaction and company image. "Effects analysis" means studying the consequences of those failures, such as severity, cause mechanisms, occurrences. So it is important to first prioritize according to how serious the consequences are, how frequently they occur and how easily they can be detected. If the failure modes are determined and accompanying effects are analyzed, actions can be taken to eliminate or reduce their occurrence and impact, starting with the highest priority failures.

Because of the systematic way of working, the FMEA study turns current knowledge and actions about the risks of failures in products and processes into transparent and traceable documentation. Many companies apply this information for continuous improvement of products and processes.

FMEA can be used during design as well as later for control, before and during ongoing operation of the process. The ideal situation is to apply an FMEA study at the earliest stage of a project and repeat and update it continuously throughout the life-cycle. The sooner failures are detected, the less expensive it is to address, eliminate or reduce them (see Figure 3.2). Nowadays there are several types of FMEAs, as outlined in Appendix B.

4.11.2 FMEA TEAM

The FMEA team should be led by a responsible design or manufacturing engineer familiar with the procedure. Team members can be design engineers, process engineers, manufacturing engineers, suppliers, logistic engineers, customers, operators, and persons responsible for maintenance. Prior to assembling the team, the responsible FMEA engineer can arrange a meeting with two or three key engineers responsible for design, quality and testing. In advance they can determine the scope of the FMEA study, elicit background reference material from previous FMEA studies and the context of the project, creating and updating function block diagrams of the system, product and/or process. Further on, they identify the team members, prepare the agenda, schedule, milestones and a preliminary overview of main item functions, failure modes and their effects is generated.

4.11.3 FMEA FORM

As an aid for performing an FMEA, a form as shown in Table 4.5 is used. It consists of four parts, each with several columns. The first part identifies the failure modes and their effects (columns 1–4); the second part the causes of the failure modes and controls (columns 5–8); the third part the prioritization (column 9) and the last part (columns 10–16) determines and assesses results of actions taken. Before the columns can be filled for each issue in the procedure outlined below, the header should be completed: All team members and the revision date must be listed (see §4.13.1). Product names and numbers (ID's) must also be detailed in the header.

4.11.4 FMEA PROCEDURE

The FMEA procedure consists of 10 steps. The numbers in brackets [] indicate the relevant column(s) of the form.

1. List the key process steps [1]. Every FMEA should detail the assumptions and ratings used. This information may come from the highest ranked items of the previously developed functional block diagram. It describes *what* the system or component is designed to do and the environment in which the system should operate. For example: The storm-surge barrier must close the tide river within 4 hours after measurement of the critical height of an increasing sea level and must open within 3 hours after measurement of the critical height of a descending sea level (as specified in functional performance specification. Include reference to source document.
2. List the potential failure mode for each process step [2]. Potential failure is the way a system, product or process may potentially fail to meet the specified objectives. The main question should be: How could this system, product or process fail to meet each customer requirement or company image? Consider potential failure modes under conditions such as operating environment (dusty and dirty or wet and dry), usage (severe circumstances), incorrect service (wrong part, mounted backward, hard to assemble). For the storm surge barrier, blocking water level is higher than the actual height.
3. List the effects of this failure mode [3]. The main question is: If the failure occurs then what are the consequences to whom? Describe the effects in terms of the exposure. The most important issues are the impacts on safety and noncompliance with law and regulations. Be aware that there is no single customer and that several stakeholders all have their own requirements. For example, the effect of a barrier

Table 4.5 The Failure Mode and Effect Analysis form

Process or Product:		Prepared by:		Page:	
Process owner:		FMEA date:		Rev:	

[1] Process step or input	[2] Potential failure mode	[3] Potential failure effects	[4] Severity	[5] Potential causes	[6] Occurrence	[7] Current controls	[8] Detection	[9] RPN	[10] Recom-mended actions	[11] Respon-sible	[12] Actions taken	[13] SEV	[14] OCC	[15] DET	[16] RPN
What is the process step or input?	In what ways can the process step or input fail?	What is the impact on the output variables once it fails?	How severe is the effect to the customer?	What causes the key input to go wrong?	How often does cause or FM occur?	What are existing controls and pro-cedures that prevent the cause or FM?	How well can you detect the cause or FM?		What are the actions for reducing occur-rence or improv-ing detec-tion?	Who is responsi-ble for the action?	Note the actions taken. Include dates of comple-tion.				
Identify failure modes and effects				Identify causes and controls				Prio	Determine and assess actions						

Table 4.6

Severity ratings for an FMEA. These can be used as a general guideline. Complete agreement is required for each project

Value	Description
10	Safety issue or regulatory violation without warning
9	Safety issue or regulatory violation with warning
8	Main function is lost or seriously degraded
7	Subfunction is reduced and customer is impacted
6	Subfunction is lost or seriously degraded
5	Subfunction is reduced and customer is impacted
4	Loss of function or appearance such that most customers would return product or stop using service
3	Loss of function or appearance that is noticed by customers but would not result in return or loss of service
2	Loss of function or appearance that is unlikely to be noticed by customers and would not result in return or loss of service
1	Little or no impact

that fails to close is that water flows over the dykes and may immerse and destroy harbor facilities.

4. List the severity effects [4] and rate them from 1 to 10 (1 = not severe; 10 = extremely severe); see Table 4.6. Ensure the team understands and agrees to the severity scale at the start. Severity measures the seriousness of the effects of a potential failure mode. A failure may impact the next component, the entire subsystem or one or all customers. For failure modes with potential for multiple effects, the highest severity rating should be selected. For example, water flowing over dykes has a severity level of 10; inability of ships to enter a harbor rates a level of 5.
5. List causes of failure [5] and occurrence [6]. Identify the causes of the failure mode or effect and rank them in the occurrence column as done for effects in column 4. This step scores the likelihood that a failure will occur (1 = highly unlikely; 10 = almost certain); see Table 4.7(a). Every imaginable failure cause should be listed concisely and completely. In the storm surge case, incorrect positions of level indicators rate an occurrence value of 5.
6. List current process control [7] and detection code [8]. Identify the control procedure or mechanism in place to detect the issue. This involves the activities that will assure the design adequacy (see §2.3.4) and measurement or detection systems for the failure cause or mechanism under consideration. These controls will detect causes and subsequent failure modes prior to production and operation, and/or will prevent the cause from occurring. Three types of control can be dis-

Table 4.7
Occurrence and detection definitions for an FMEA

(a) Example for definition of occurrence

Value	Description
10	1 in 2
9	1 in 10
8	1 in 50
7	1 in 250
6	1 in 1 000
5	1 in 5 000
4	1 in 10 000
3	1 in 50 000
2	1 in 250 000
1	1 in 1 million

(b) Example for definition of detection values. The description indicates the chance of detection of the failure

Value	Description
10	uncertain
9	very remote chance
8	remote chance
7	very low chance
6	low chance
5	moderate chance
4	moderately high chance
3	high chance
2	very high chance
1	almost certain

tinguished:

a. Prevention
b. Detection of cause mechanism
c. Detection of failure mode.

One can for instance review the specifications thoroughly or measure the condition of the surge barriers bearings regularly. The detection scale is *larger* for items that are more *unlikely* to be detected, with 10 as the least likely; see Table 4.7(b). The greater the probability of detection, the lower the rating. The water level sensor of the barrier is scheduled to be checked daily and so the detection rating for water level is 1.

7. List the Risk Priority Number (RPN) [9]. The RPN is a multiplication of severity, occurrence, and detection ratings:

$$RPN = Severity \times Occurrence \times Detection$$

This number will be used to identify where the team should focus first. The RPN range runs from 1 to 1000. The team must make efforts to reduce higher RPNs through corrective action. General guideline is that action is required for all failure modes with RPNs over 100.

8. List recommended actions [10]. Sort by descending RPN numbers and identify the most critical issues. The team must decide where to focus first. Only design revisions can result in changes to the severity ranking. Examples of recommended actions are reliability testing and revisions of test plans and design.
9. List responsibility and target completion date [11]. Assign specific actions to responsible persons.

10. List the action results [12-16]. Once actions have been completed, re-score the occurrence and detection. In most cases we will not change the severity score unless the customer decides the potential problem is not an important issue. Unless the failure mode has been eliminated, severity should not change.

To apply an FMEA successfully ask yourself and team members a lot of questions, such as: what would happen if ... or can this or that happen ...? Provide for mutual understanding about functions of system, product or process. FMEA processes can fail as well if members of the team are not familiar with the principles and tools or try to rush through the process. The preparation stage of the FMEA study should be worked out carefully, consistently, transparently and traceably so that every team member and other stakeholders can understand the results of each step.

4.12 RISK MANAGEMENT

Risk is a very general term, and its meaning is not always clearly defined. For development projects, it is necessary to consider two aspects of risks R, namely the probability P_r and the magnitude of the consequences C. More formally defined:

$$R = P_r \times C$$

where R, P_r, C are all on a relative 0 to 1 scale. For probability, such a relative scale is easily grasped; for consequences, it is more difficult. One can for instance define 1 as complete loss of the total project budget. It is essential to develop a clear definition of each risk and communicate it across the organization.

An event that has only a small chance of occurring but presents catastrophic consequences represents a high risk as does an event that is highly probable but may produce moderate consequences; see Figure 4.15. If a project involves multiple risks, the overall project risk can be calculated by focusing on the chances of success rather than on the chances of failure as shown in the following equation: If in a project multiple risks occur, then the overall project risk can be calculated by focusing on the chance on success, instead of on the chance of the chance of the failure:

§3.13: Risk thinking

$$P_{\text{success}} = (1 - P_{r1}) \cdot (1 - P_{r2}) \ldots (1 - P_{rn})$$

And thus:

$$\Rightarrow P_{r\,\text{total}} = 1 - P_{\text{success}}$$

Of course, although this looks like exact analysis, the numbers representing the risks are based on estimates. Therefore, risk analysis should be considered a qualitative affair.

For risk *management* defined as keeping the risks in the project within acceptable limits, we discuss the following tools in separate subsections:

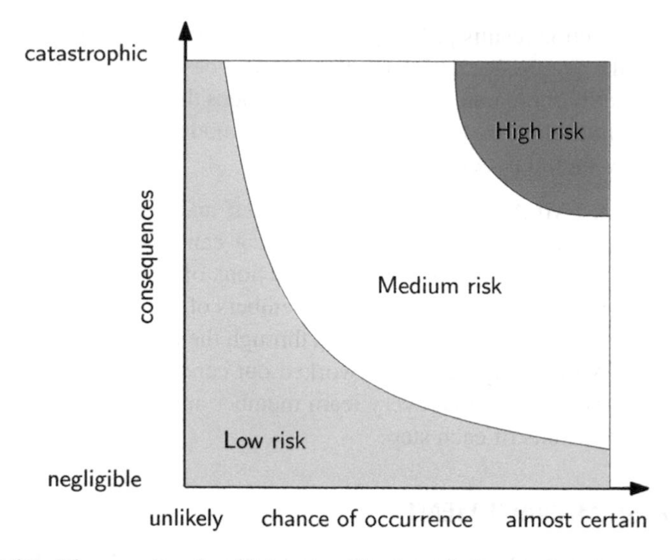

Figure 4.15 The severity of a risk is a combination of chance of occurrence and consequences.

- Decision tree, to quantify risks when making a decision
- Risk register, to track the number and magnitude of risks in a project

4.12.1 DECISION TREE

In its simplest form (Figure 4.16), a decision tree maps out the expected decisions to be made and represents the alternatives as branches. As example, Figure 4.16 shows three alternatives that can be taken: upgrading an existing system; creating an entirely new design, or buying a system in an alliance with another organization. For each of the alternatives, there are three possible outcomes once a prototype is available:

1. Performance is according to the requirements (OK).
2. Performance is not ok (NOK), but a minor redesign will suffice.
3. Performance is not ok and a major redesign is required.

A second use of the decision tree is to quantify each of the courses, that is, to analyze the probability of each of the alternatives and quantify the time and money involved in each of the tracks. This is shown in Figure 4.17. Probabilities branching from one point have to add to unity. For the first branch the meaning of the probabilities can be determined based on how feasible is the track expected to be. This can be done by considering whether suitable partners for an alliance are available (third alternative) and experience with the present system (first alternative). From these estimates, based on past

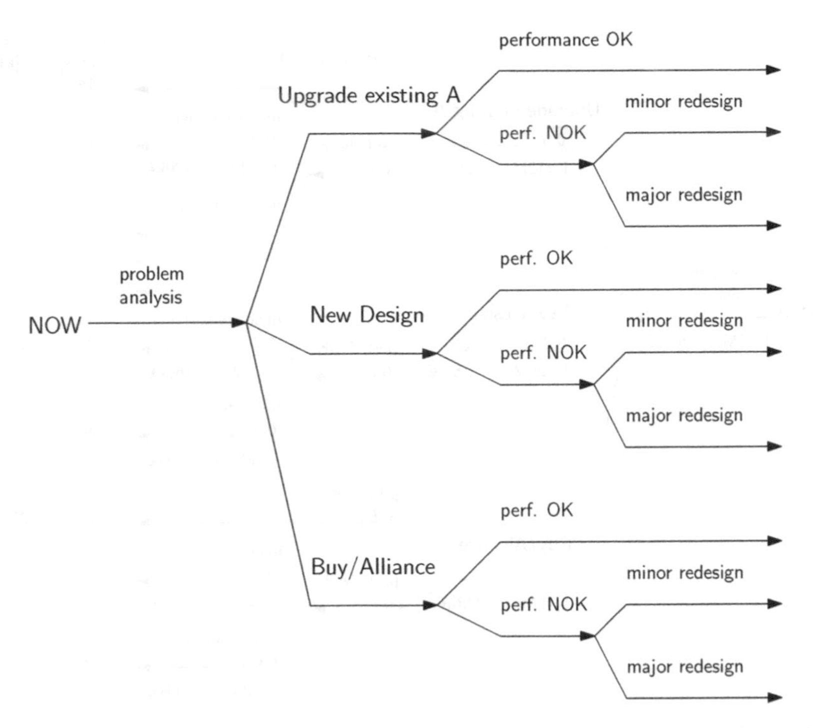

Figure 4.16 A decision tree in its simplest form to map the possible course of a project.

experience, interviews, and preliminary modeling, the time and cost of each of the branches can be calculated as shown by Figure 4.17.

The decision tree is particular useful when a bid calculation for a project has to be made. Referring to the right side of Figure 4.17, one can see the minimum and maximum cost (C) and running time (T): $1050 \leq C \leq 3050$ [k€] and $10 \leq T \leq 30$ [months], respectively. Depending on whether the calculation has to be done safe on the timing or the financial side, the choice can be made either to bid for 30 months, or for a 3 M€ offering, possibly with a profit and/or margin added. However, other issues come into play as well: how important is this project for the continuity of the organization? Is there some prestige to gain? Do we have a long-term relationship with the customer. If so, it may be worthwhile to bid with more risk at a lower price. Here, the probabilities can be used.

Every alternative in the decision tree can be assigned a probability, as shown in Figure 4.17. Multiplying the probabilities along a route (for example the top route from NOW, to upgrade existing A [0.5], performance OK [0.3], resulting in a probability of 0.15 for this branch) gives the estimated probability of this variant occurring; see Table 4.8. Even a mean expected value can be determined. This value should be considered with caution because it is based on

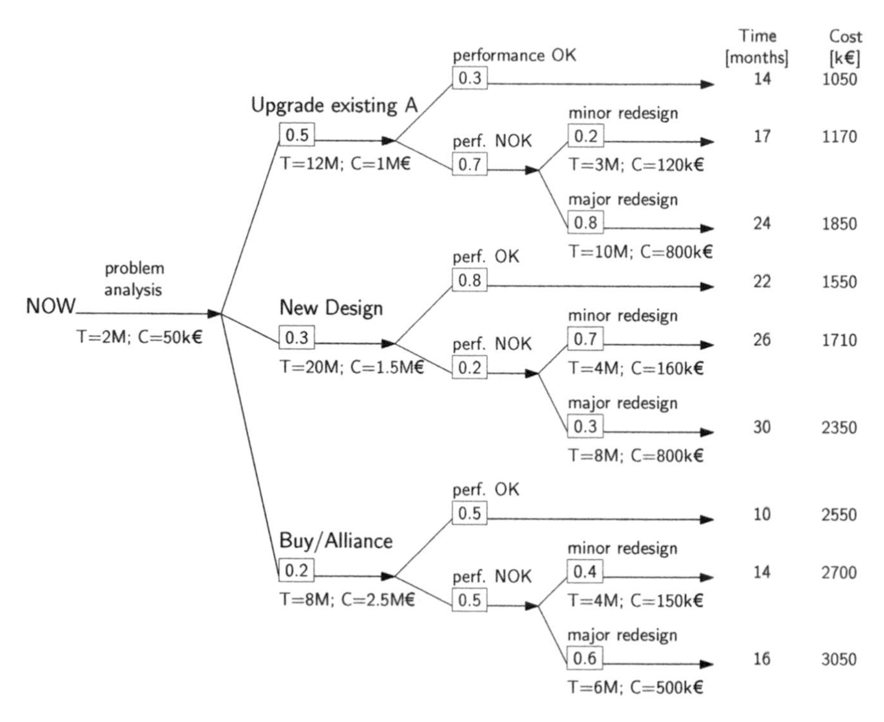

Figure 4.17 Decision tree for analysis of bid calculations. The cost (C) and time (T) of each step have been added, as well as the probability that an alternative will occur (boxed number). At right, each of the alternatives is assigned a total running time and cost by summing the time and cost along the track of the alternative.

probabilistics. Only when several projects a year are treated this way, can one use this to make a bid. In other cases, an analysis like this does reveal margins and opportunities. A table like Table 4.8 summarizes the minima, maxima, means and expected value and is a good basis for decision making.

4.12.2 RISK REGISTER

Where the decision tree is useful in the project planning phase, the risk register that we treat in this section is useful while a project is underway. Some of the elements in the decision tree reappear in the risk register and some new elements are added.

For each identified risk the following is listed; most are self explanatory, if they are not, short descriptions are given:

1. Identification
2. Short description of the identified risk
3. Owner
4. Start date

Table 4.8

Evaluating the decision tree quantitatively. The columns Prob, Time and Cost are derived directly from the decision tree in Figure 4.17. The columns Time$_w$ and Cost$_w$ are the weighted time and cost, respectively, calculated by multiplying the time and cost with the probability. The expected values are the summed time and cost over all variants

Variant	Redesign?	Prob [-]	Time [Months]	Cost [k€]	Time$_w$ [Months]	Cost$_w$ [k€]
Upgrade A	none	0.15	14	1050	2.1	157.5
	minor	0.07	17	1170	1.19	81.9
	major	0.28	24	1850	6.72	518.
New design	none	0.24	22	1550	5.28	372.
	minor	0.042	26	1710	1.092	71.8
	major	0.018	30	2350	0.54	42.3
Buy/alliance	none	0.1	10	2550	1.	255.
	minor	0.04	14	2700	0.56	108.
	major	0.06	16	3050	0.96	183.
Expected values					19.4	1789.5

5. Relief date (the last moment that action has to be taken)
6. Cost impact (the total cost when the risk becomes reality)
7. Probability
8. Weighted cost (as in §4.12.1, Cost impact × Probability)
9. Status/trend (status describes the actual situation. If the risk relates to mass or performance, one can indicate "too high" or "reducing")
10. Strategy, including action plan and cost of action plan. The cost of the action plan has to be *explicitly* compared to the weighted and total cost of the identified risk. Also the *total* cost of the action plan and total risk costs have to be compared.

A risk register should be regularly updated and discussed at the regular project meetings. If the project meeting takes place weekly, the risk register is updated and discussed weekly. Table 4.9 shows a risk register used in the solar team project.

4.13 DOCUMENTATION AND REVIEWING

In many books on systems engineering, *documentation* plays a central role. In this book, we devote only one section on this subject. Nevertheless, every systems engineer should be aware that thinking, brainstorming, drawing, discussing and commenting are easy, but putting facts on paper (or in digital form) and getting such a document accepted by the relevant stakeholders is sometimes difficult. A written record can reveal inconsistencies. As noted by

Table 4.9

Risk register as risk inventory and management tool. *Ow* is Owner (indicated with initials); dates are shown in the short ISO notation (YYMMDD). The T/s (=trend/status) column uses the following symbols: *o*: stable, +: increasing, –: reducing. For the strategy type: *bac* means backup scenario, *mon* is monitoring, *sim* is simulating

ID	Identified risk	Ow	Dates		Cost	P	$Cost_w$	T/	Strategies		
			start	relief	€	%	€	s	type	description	cost
1	Panel defect	F.	110315	110615	9000	15	1350	o	bac	Backup panel	9000
2	Scrutineering	B.	111010	111016	3000	12	360	o	sim	Do test scrutiny	150
3	Extra media costs	A.	110501	111031	9000	25	2250	+	mon	Find sponsors	1000
4											
Total					21000		3960				10150

Eising [2007], "Check whether it can withstand writing down". Also, a document requires commitment because it indicates that the writer and reviewer stands by the information.

§3.11.3: Organization life-cycle thinking

§3.10: Organizational thinking

Another purpose of documentation is to decouple development processes. This was already mentioned in §2.3 related to creating the system design. Thus the documentation structure is the reflection of the process, and therefore as a reflection on a meta level the close relation between treating the process and documentation in many systems engineering texts is easily understood.

In this section, general documentation rules are given (§4.13.1), then we provide specific guidelines for the documents in the systems engineering process, as presented in §2.3 and Figure 2.4. The section ends with guidelines for reviewing documents (§4.13.3).

4.13.1 GENERAL DOCUMENTATION GUIDELINES

Every document should be traceable. This means that the first page(s) should clearly indicate:

- Title and short summary
- Date
- Version (and record the version and update history in the document)
- Author (name and role in organization)
- Owner (name and role in organization)
- Status including whether draft or concept is accepted and change control (accepted or not accepted and responsible party)
- Documents on which this document is based
- Distribution list

Avoid duplicating information in a document. This is a general rule in information technology. Therefore, state the information in one place and use references in other places.

Be consistent in the use of terms and units, and use them explicitly. A glossary in the document or even a global glossary for the project or organization may be needed and this is a systems engineering task.

It is wise for an organization to have a template, like the one shown in Appendix C, documents are created easily and usually and so that authors are guided in filling in the details. An author of a document is namely a (system) engineer, not a writer. Note that such a template is optimized not for beauty but for effectivity and efficiency.

The organization also has to define archiving and naming conventions. There are numerous methods and philosophies; this book is not the place to elaborate on them. In fact, the particular convention chosen is of less importance than the fact that if one is chosen it has to be clearly communicated and all people involved must adhere to it. Two remarks are in order:

- Follow the ISO format: YYYYMMDD as it allows for sorting (also the short ISO date format may be used: YYMMDD).
- Make sure that archiving is done and regular backups are made.

Depending on the way the documents are created and updated, it can be wise to use a computerized versioning system. There are several available, like Apache SVN, which can also be used for version control of software code generation.

Returning to the remark that engineers are not authors, documents have to be readable and use proper language. Also, diagrams should be of good quality. The general rule should be that a document is written for the reader(s), not for the author.

4.13.2 DOCUMENT CONTENTS

In this section, we will give a short description of the contents of SE documents. The consequences of the system engineering process as outlined in Chapter 2 are that there is an alternating process of interpretation of the need resulting in requirements and interpretation of the requirements resulting in a design (see Figures 2.3 and 2.4). Thus, apart from the customer wish, there is an alternating sequence of design requirements, etc. Therefore we will treat these two types of documents and other documents in separate subsections below.

4.13.2.1 Requirements Document

A requirements document like a system requirement specification (SRS) or a subsystem or element requirement specification (ERS) describes what performance level the system has to achieve, without describing the solution (describing the *what*, without the *how*). It is, as said above, an interpretation of the need such that:

Table 4.10
Checklist for a requirement specification document, based on Eger et al. 2013

A requirement specification document

- Treats all life-cycle phases of the system (see §3.11).
- Treats the performance of the system, referring to the functions.
- Differentiates between *must haves* and *likes*.
- Indicates functions that are needed, but are not part of the system under design (derived functions).
- Can be used to test the system under design and thus measurable criteria are required.
- Is not aimed at one solution (in other words: the solution is not contained in the requirements specification[7]);
- Is feasible and supported with tests, experiments, and experience etc.

- The customer can verify that what will be delivered is according to his wishes
- The builder knows the goal of his development effort

A checklist for a requirement specification is given in Table 4.10.

The requirements specification is written by the team that has to create the (sub-)system. The system requirements are written by the system designers, based on the customer wish; the subsystem requirements are written by the subsystem design team, based on the system design specification. This is needed to ensure that the design loop (Figure 2.3) is closed. The requirements are interpreted by the persons who must create the solution. The review process (§4.13.3) is an essential part of closing the loop as well.

4.13.2.2 Design Document

The design document like a system design specification (SDS) or subsystem or element design specification (EDS) describes the solution chosen to achieve the performance (the *how*).

§3.5: Operational thinking

On the system level the document is an identification of the subsystems with the definition of the interfaces between them. Also, the main system concept has to be described and supported. How this is done depends on the industry and company. The design document can be an artist's impression, a very quantitative description, or a combination. Also, the items that have very

[7] In contrast with a design document page 78.

long development or production times—the so called *long-lead items*—have to be identified and detailed, so that their production can start as soon as possible. This has the consequence that their design cannot be changed. Note that A3AOs (§4.10) are of use here too.

On subsystem and lower levels the documents become more and more specific. Also, there will be more mono-disciplinary documents like mechanical CAD plans, electrical schematics, and simulation codes, that are part of the documentation for the subsystem. Nevertheless, all the documents have to comply to the subsystem requirements, and this has to be checked both by the author of the design document and in a review (see §4.13.3).

4.13.2.3 Other Documents

While the requirements document and design document have a close relation to the system design process as presented in Chapter 2, and more specifically in Figure 2.4, during the course of the system design process more documents are produced:

Investigation reports: present the results of diverse investigations:

- Possible concepts
- Performance of competitors' products
- Possible causes of a problem
- Areas for improvement of the current system(s)
- Design and results of experiments

Test protocols: define a test procedure. These are closely related to the requirements documents in the sense that the combined test protocols have to show that the requirements are met (note the link already made in Figure 2.8). Important test protocols are the acceptance tests that are conducted with representatives of the customer just before or just after delivery, and that have to be run before payment is completed. These protocols are the official validation of the system (see §2.3.4).

Test reports: describe the set-up of tests and results. An evaluation of the test, the results and the test method should be part of each test report. Test reports often detail components and subsystems.

These documents can serve as supporting information for other documents, like requirements documents and design documents. If so, they should be referenced from these requirements and/or design documents.

4.13.3 REVIEWING DOCUMENTS

As Figure 2.3 shows, designing is a cyclic process. Documents are created to describe requirements, designs, experiments etc. A document is the condensation of the process shown in Figure 2.3, and can be used for building a next step in the system design process (Figure 2.4). This requires the document to

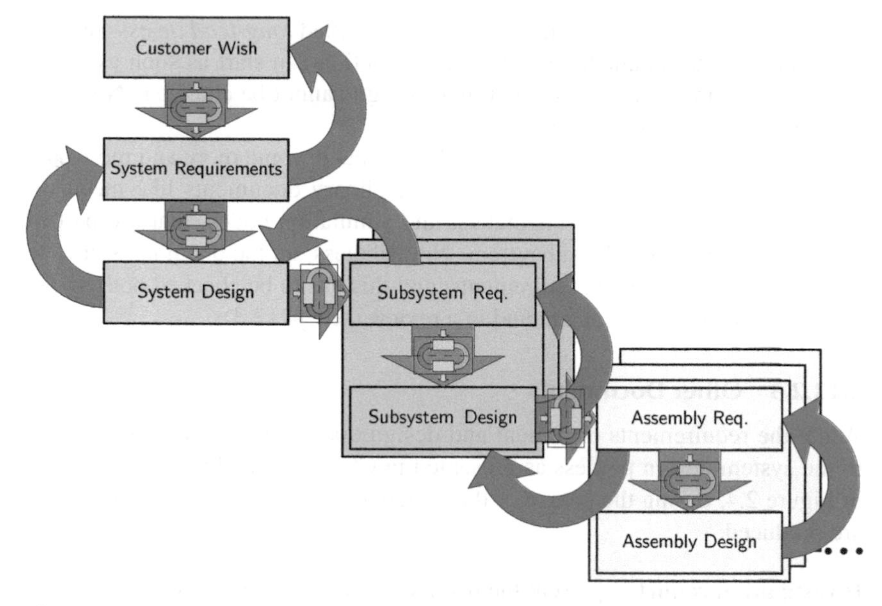

Figure 4.18 The systems engineering process shown with the reviews in place. The reviews act as feedback throughout the process.

be checked, accepted and stable. The checking and accepting is done through reviewing.

A classical review is done by people sitting in a room and going through the document page by page. The people involved are:

- Author
- Client (the one using the unit; this does not have to be the end customer, but can be someone higher up in the hierarchical chain; see Figures 2.2 and 2.4)
- Builder (e.g. the one who has to realize the unit)
- Supervisor of the builder
- One or more experts
- Representatives of (sub)systems that interface with the unit
- If needed a facilitator and/or a representative from the systems engineering department

The experts can be people who have designed or built comparable units or who have written the requirements. Reviews can be done online using tracking systems. However, this does eliminate the interaction between the people, and removes triggers that exist when people are together in a room.

What should be checked? Reviews should be seen as feedback in the design process, as shown in Figure 4.18 and as a communication aspect of creating mutual agreement on language, performance and in case of design documents, solutions.

Therefore the following should be checked:

- Consistency
  - In terms and language within the document and with documents higher in the chain
  - In performance (to ensure compliance with (sub)system requirements and budgets)
  - Of interface specifications
- Compliance
  - With company values and mission
  - With system concept
- Consequences
  - For other (sub)systems on the same hierarchical level
  - For subsystems lower in hierarchy
  - For project planning
  - For reuse of (sub)systems and components

The result of a review can be that the document is accepted, is accepted with some modifications, or remains a concept. In case of acceptance, the document is put under chance control and cannot be altered without notification and consequences because many stakeholders have based their work upon it. If changes are needed, a new review with the same committee is required.

4.13.4 OPEN ISSUES AND DECISIONS

No matter how thorough the documentation and reviewing process has been, the reality will always reveal issues not treated earlier. That is the nature of design [Schön and Bennet, 1996]. Also, present day development processes are done *concurrently*, meaning that different phases run partly in parallel. Thus, not all decisions have been made before the next step is initiated. This relates to the uncertainty mentioned in §3.1. One way to deal with this situation is to continuously update the documentation. While this is the best way, it would take too much time. A more practical approach is to introduce two lists under control of the systems engineer or systems engineering team:

- Open issues
- Decisions

The open issues list contains items that need to be addressed, including a short description, priority, problem owner and parties who need to be involved.

The open issues can then be addressed systematically based on their priorities and decisions should result. A list of decisions should be maintained and each entry should show the expected result and the parties who will be affected. It is advisable to link decision lists to meeting minutes.

The open issue and decision lists should be managed by the SE and should be accessible to all members of the design team. Completion dates and owners of all list items should be shown.

4.14 MODELING AND SIMULATION

§3.7: Scientific thinking

§3.3: Feedback thinking

§3.2: Dynamic thinking

A symbolic model in the form of equations or statements can be simulated to see how the model behaves. Dynamic models of engineering systems (i.e., describing the behavior of the model over time) mostly consist of *differential equations* (ODEs), possibly extended with algebraic constraint equations. The generic form is:

$$\dot{x} = f(x,u,t) \tag{4.1}$$

$$y = g(x,u,t) \tag{4.2}$$

where Equation (4.1) represents the set of differential equations, with x representing so-called *state variables*, and u the input variables; Equation (4.2) represents the output equations needed to compute other variables in the system (y), which are not necessary to compute the state variables x. The differential equations may be implicit, implying some algebraic constraint equations are present; hence the term DAE (Differential and Algebraic Equations); Equation (4.1) becomes:

$$\dot{x} = f(x,\dot{x},u,t) \tag{4.3}$$

These models can be simulated on a computer using standard numerical integration methods, which are basically *approximations* of the mathematical integration functionality. The independent variable *time* will be discretized, i.e., only at discrete points t_k the model is computed (otherwise we would have an infinite number of time points during a simulation). Further, the numerical integration formulas are approximations of the theoretical integration process. This implies that different numerical integration methods exist, each having its own class of models for which it is suitable in terms of accuracy and simulation speed. The choice of the best usable numerical integration method can be tricky, although for normal models, a contemporary simulation tool like Simulink or 20-sim provides suitable integration methods.

During a simulation, the result of the model computation (Equation (4.1)) is used to compute the next value of the states x, indicated as x_{k+1}. After assigning the initial values to the states, successive values of x_k and y_k are calculated successively by evaluating the model equations and applying the numerical integration method alternatingly (right loop in Figure 4.19), until the specified end time is reached. Note that most numerical integration methods need calculation of the model equations (4.1) as part of the numerical integration process (the left loop in Figure 4.19).

Once the simulation is done or often during the simulation, the results can be visualized in many ways. Graphs and plots are effective but modern computer tools include animators that can show a simplified version of a product in motion. Often a combination of a mechanical CAD tool and a simulation program can deliver a rich visualization.

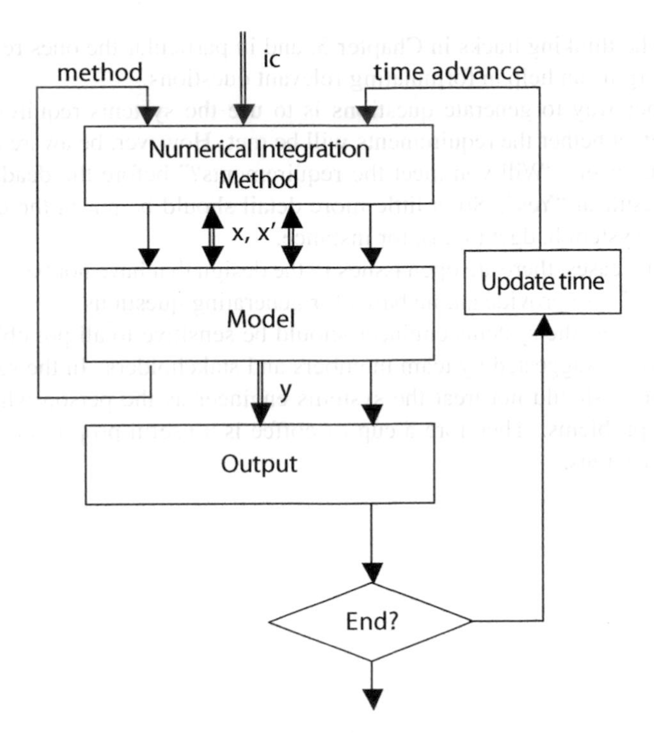

Figure 4.19 The simulation process [Broenink, 1990].

4.15 QUESTION GENERATOR

When the project is through the initial phase (the system column in Figure 2.4), and the team size has increased, the role of the systems engineer changes, as discussed in §5: he becomes more of a watchdog and knowledge keeper, and also must probe the decisions made and the impact they have on the system as a whole. Walking around, asking questions, and talking over a cup of coffee may be useful. The questions should not be random, but directed to the system function (both normal operation and exceptions). There are several ways to generate questions.

In [Muller, 2004a], a WWHWWW (why, what, how, who, when, where) question generator is proposed to sample the design space. The generator uses a simple rule:

§3.4: Specific-generic thinking

How about the `<characteristic>`
of the `<component>`
when performing `<function>`?

§3.8: Decomposition–composition thinking

Based on the background of the systems engineer, there are many issues to ask detailed questions about. These can relate to mechanical construction issues or electronic matters and also physics that may limit the operation of the

§3.9: Hierarchical thinking

S.U.D. The thinking tracks in Chapter 3, and in particular the ones referred to in the margin can help in formulating relevant questions.

Another way to generate questions is to use the systems requirements to investigate whether the requirements will be met. However, be aware that asking an engineer: "Will you meet the requirements?" before the deadline will always result in "Yes". So, a little more detail should be put in the question, using the system budget (§4.8), for instance.

In many cases, there are open issues in the design that have not been detailed (§4.13.4). These provide ample bases for generating questions.

In all cases, the systems engineer should be sensitive to all possible issues and solutions suggested by team members and stakeholders. In the same way the engineer should not treat the systems engineer as the person who has to solve all problems. Therefore a cup of coffee is a useful prop to carry while asking questions.

5 The Systems Engineer at Work

With the three compartments of the systems designer and engineering toolbox filled (the process, the thinking tracks and the tools), it is time to look at how the system designer at work operates.

The blue parts in Figures 2.4 and 2.6 show the visible work and output of the systems engineer. In the design phase, it is the systems engineer's responsibility to develop the customer's wish into the requirements and then into a realizable system design as described in §2.3.1. It is here where the systems engineer, system designer, or system architect, (we will use the systems engineer designation in this chapter) has to get input from the customer and other stakeholders. Depending on the type of industry, organization, and product, this means:

- Direct contact with the customer through interviews and/or customer visits
- Market research through interviews with experts and observation [Eger et al., 2013];
- Acquiring input from the marketing department

This input has to be reworked and balanced with other stakeholders' requirements into a balanced set of key drivers [Heemels et al., 2006].

§4.9: FunKey architecting

In parallel, the systems engineer has to be up to date with technology inside and outside the organization and know what can be achieved by the engineers.

Both have to be combined into an architecture: a set of representations (like the ones treated in Chapter 4, but others may be needed) that describes the system to be designed, so that the engineers know what they have to do, and the customers and other stakeholders' know what they can expect. In fact, the architecture is an interface between the engineering achievements and the stakeholders' expectations, as shown in Figure 5.1 (such an interface was also described when introducing the key drivers in §4.9 and Figure 4.12). The architecture description should be a combination of functional, physical and quantified models, as already shown in the A3 architecture overviews (§4.10).

§4.7: Architectures

§4.10: A3 Architecture overviews

An important part of the systems engineer's task is balancing the stakeholders' wishes and requirements and making trade-offs. Also, as stated in Chapter 2, the definition of subsystems and the interfaces is the systems engineer's core business. In some cases the fundamentals of the systems, including the physical principles, have to be chosen.

In the integration, verification and validation phase, the systems engineer is involved as knowledge holder; as the designer of the system, he is most

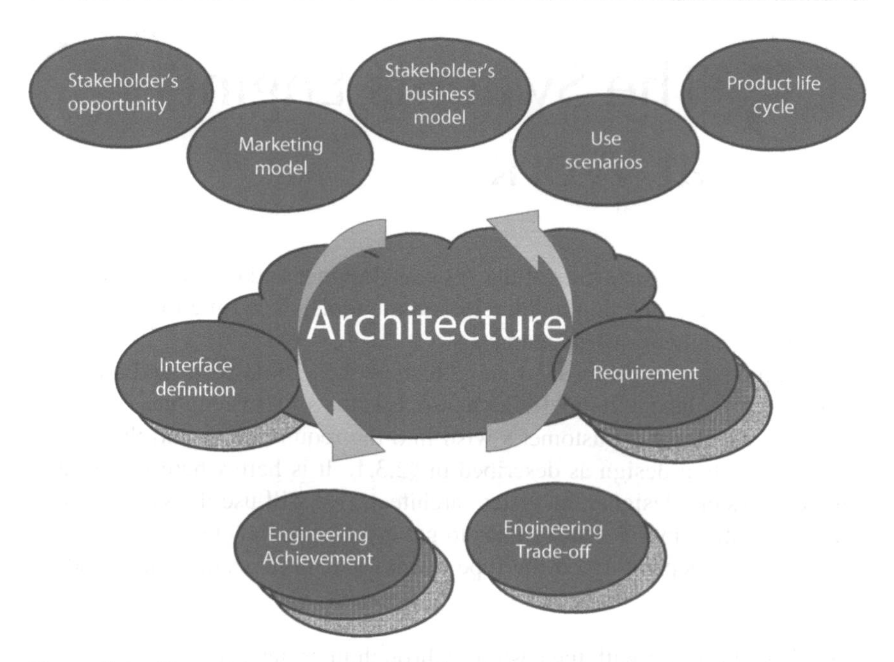

Figure 5.1 The architecture as interface between stakeholder expectations and engineering achievements [Bonnema, 2014; Bonnema et al., 2010].

aware of the system's composition, and can thus deal with issues regarding interactions between components and assemblies. Also, when subsystems and ultimately the entire system, are verified against the requirements, the systems engineer should be involved to make sure that meaningful results for the customer (and end-users) are gathered. This means that the systems engineer must have a good image of what creates value for the customer. *Validation* (§2.3.4) is best done by or in cooperation with the systems engineer.

§4.15: Question generator

Does this mean that the systems engineer has little to do between these activities? Quite the contrary. He is busy safeguarding the concept and monitoring the design process. In particular, he should pay attention to the following issues:

§4.12: Risk management

§4.8: System Budgets

- Is the concept still maintained?
- Will the requirements be met?
- Are the interfaces implemented and used and not circumvented?
- Identify, address and solve risks and problems with the designers involved,
- Continuously monitor the balance in the system.

A systems engineer should be well educated in a broad set of disciplines. In general there is one home field, with several adjacent fields where he is comfortable. In addition; he should communicate easily and frequently with

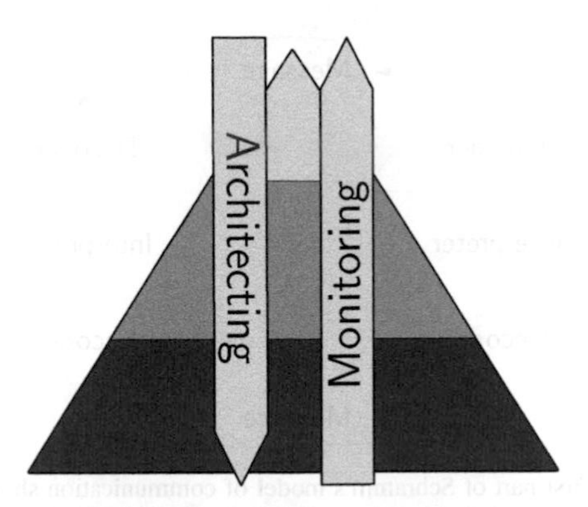

Figure 5.2 Architecting and monitoring the development process in the Muller pyramid shown in Figure 2.2 [Bonnema, 2008].

engineers and non engineers alike. The relation between the design and architecting process and the monitoring process is shown in Figure 5.2. The activities of architecting and monitoring complement each other.

An important observation is that the systems engineer should learn to live with incomplete and uncertain information. Using intuition and experience, he has to make decisions in the absence of certainty; see Figure 3.2.

5.1 COMMUNICATION IN SYSTEMS ENGINEERING

An underlying aspect of the systems engineer's job is to facilitate communication among disciplines. This aspect has been just below the waterline throughout this book. Here it makes sense to explicitly address it.

Modern-day systems are created by the joint effort of many people with different backgrounds: mechanical engineers, software engineers, ergonomics experts, and marketeers. It is essential for the systems engineer to acknowledge that each of these developers has his own background, way of working and expressing himself. The systems engineer then has to make sure that all efforts are directed to delivering one well-functioning and balanced system that will generate value for the customer(s).

It is a challenge to have these developers work together and use a common frame of reference (the baseline we talked about earlier). A software engineer may prefer a baseline expressed in UML (Unified Modeling Language) or SysML. A mechanical engineer prefers geometry-related models, such as mechanical CAD models. Proper cooperation among them requires a common field of reference [Schramm, 1954]. [Clark and Brennan, 1991] use the term "common ground" that is also shown in Figure 5.3(b).

§4.10: A3 Architecture overviews

In order to close the communication loop (Figure 5.3), it is therefore important that the systems engineer uses models and representations that are understandable by a wide range of stakeholders including representatives of the

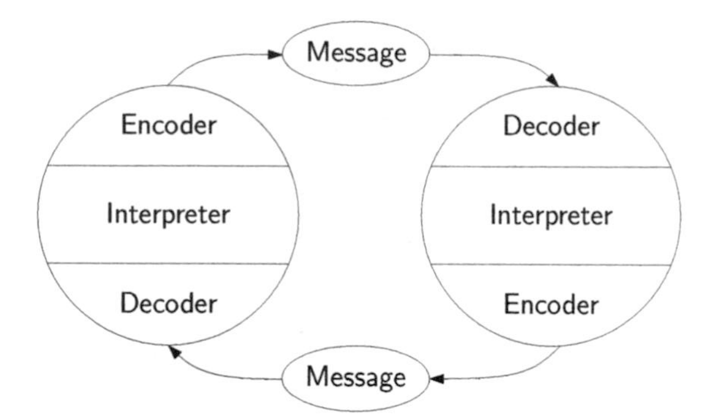

(a) First part of Schramm's model of communication showing feedback and the encoder and decoder at the sender's and receiver's sides.

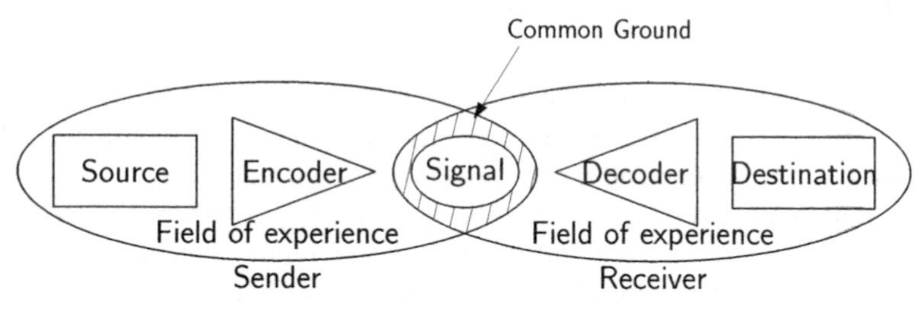

(b) Second part of Schramm's model of communication showing fields of experience of both the sender and the receiver. The hashed area represents the common ground [Clark and Brennan, 1991].

Figure 5.3 Schramm's model of communication [1954].

customer and users. The A3 Architecture Overviews (§4.10) provide widely understandable representations. Other combinations of representations may be used as well, depending on the culture and background of the developers.

In general we can state that the handling of and interacting with the chosen representations should not use too much effort of the stakeholders. Therefore, we recommend the following (as concluded in [Bonnema, 2014]):

- Multiview: use various formalisms to connect to the diverse stakeholders involved. Also note the IEEE1471 and ISO/IEC/IEEE42010 multiple view requirements.
- Simplify as much as possible. This requires a significant effort from the architect to find the essence. Different representations may be required for different discussions, as the essential components may differ.
- Connect to reality. The representations should resemble the system under design. An abstract boxology is not enough. At first glance the representation of an MRI system should be different from that for a wafer stepper.

§3.4: Specific-generic thinking

§3.5: Operational thinking

- Use the tripod of functional-physical-quantification in all representations. Functional represents the abstract "what", physical represents the concrete "how" and "where", while quantification addresses the "how much".
- Use feedback. An architecture evolves and is developed by involving diverse stakeholders. By presenting intermediate results, tracking the reactions, and improving the architecture, the architecture develops in a spiral way (see Figures 4.8, 4.14 and 5.3)). Time boxing can be useful here.
- Use simple models. To support the quantification, use numbers based on reality. These can be derived with often simple models that nevertheless contain the essence. While an error margin has to be taken into account, such models and numbers are accurate enough for the architecture. When detail design begins, more precise models have to (and will be) developed.
- Communicate! This may well be the most important aspect. The architecture has to be communicated as much as possible (Figure 5.1), within development, marketing and other departments, just as input for the architecture is required from all these stakeholders. In fact, the architecture is the main interface between them: it shows the benefit from engineering achievements to marketing and the relationship of marketing opportunities to development.

§4.10: A3 Architecture overviews

§3.3: Feedback thinking

§3.6: Scales thinking

§4.8: System Budgets

§4.10: A3 Architecture overviews

5.2 THE SYSTEMS ENGINEER AND THE PROJECT MANAGER

A second issue that requires attention is the distinction between two roles in leading a development project:

- The project manager leads the project from a people, time and money perspective.
- The systems engineer leads the project from a technical and risk perspective.

This book has focused on the system engineering perspective, although some project management aspects have been touched (for instance in §3.10, 3.11.3 and 3.13). Depending on the size and complexity of a project, the two roles can be fulfilled by one person, or may be divided between two or more persons. The project manager has the power to force decisions; the systems engineer can only exert influence.

A close cooperation between the project manager and systems engineer is required for a succesful project that meets financial, timing, technical and customer satisfaction criteria.

Appendices

A TRIZ[1]

A.1 SHORT INTRODUCTION TO TRIZ

Altshuller and his team, investigated hundreds of thousands of patents looking for patterns in the inventions. This resulted in a method (or a set of tools) that can be used to solve various technical problems, hence the name Theory of Inventive Problem Solving (in Russian, the acronym is TRIZ). The common pattern in all of TRIZ is (Figure A.1), that a specific problem is translated into a generalized problem. TRIZ then provides principles and patterns for finding a generalized solution. The final step is to translate this generalized solution to the specific case at hand.

One of the most popular tools of TRIZ is the list of innovative principles (IPs) combined with the contradiction matrix. Altshuller found that many patents resulted from solving contradictions of the form: when <property A> improves, <property B> deteriorates. The properties in the specific problem can be connected to 39 parameters of a technical system [Altshuller, 1997]. Examples of these parameters are weight of moving object; speed; and reliability.

Based on the information in the patents researched, a limited number of ways to solve contradictions was found: the 40 innovative principles (IPs); see Table A.2. Thus there are, according to TRIZ, only a limited number of parameters (39) that constitute contradictions, and a limited number of solving principles (40). Therefore, the contradiction matrix was created (Table A.1 shows a part of this contradiction matrix; the full contradiction matrix can be found in Altshuller [1997]). Any possible contradiction of the form: when <technical parameter X> improves, <technical parameter Y> deteriorates, can be found in the contradiction matrix. The numbers mentioned in the cell indexed by X and Y point to one to four promising innovative principles. For example, the speed (parameter 9) of a machine has to be increased. However, this leads to an unwanted increase of forces (parameter 10). Table A.1 suggests using Innovative Principles 13, 28, 15 and 19. These are listed up in Table A.2.

Based on the work described in Ivashkov and Souchkov [2004], the priorities of TRIZ principles were determined. This was done to find the most appropriate principles to *improve* a technical parameter (Section A.1.1) and for *reducing the negative impact* on a technical parameter (Section A.1.2). The former is based on the work Ivashkov and Souchkov, 2004.

[1] This appendix has been published [Bonnema, 2011].

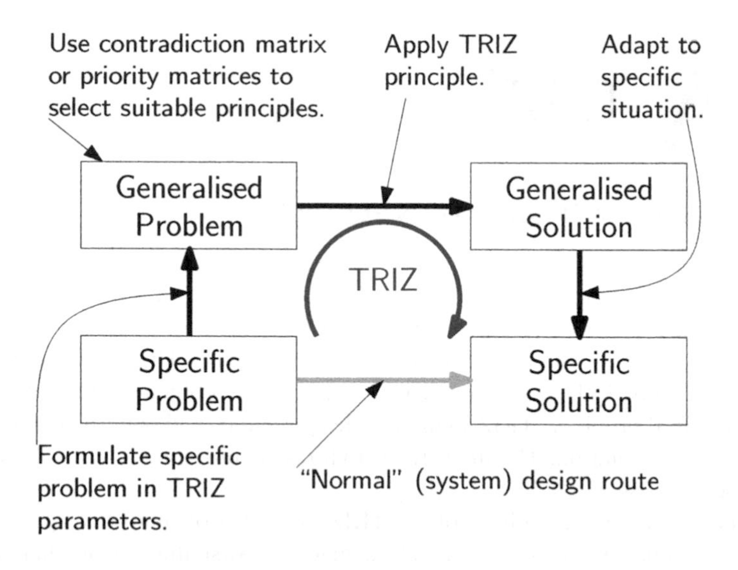

Figure A.1 The TRIZ approach to design and problem solving. By taking a "detour" via a generalised problem and solution, solving problems is simplified.

Table A.1

Part of the contradiction matrix used in TRIZ [Altshuller, 1997]. Only the segment for the technical parameters 9 through 12 is shown. In the cells, the corresponding innovative principles are shown. The numbers refer to the principles listed in Table A.2

Improving		Deteriorating			
		9	10	11	12
9	Speed		13, 28, 15, 19	6, 18, 38, 40	35, 15, 18, 34
10	Force	13, 28, 15, 12		18, 21, 11	10, 35, 40, 34
11	Tension/Pressure	6, 35, 36	36, 35, 21		35, 4, 15, 10
12	Shape	35, 15, 34, 18	35, 10, 37, 40	34, 15, 10, 14	

Please note that the body of knowledge in the TRIZ community is large and expanding rapidly. For more information, a further introduction and latest findings, please refer to `http://www.triz-journal.com/`

A.1.1 POSITIVE PRIORITY MATRIX, PM^+

For the positive priority matrix PM+, the number of occurrences of an innovative principle, IP, on a row in the TRIZ contradiction matrix (Table A.1 and Alt-

Table A.2
The 40 principles identified in Altshuller [1997]

#	Principle	#	Principle
1	Segmentation	21	Rushing through
2	Extraction	22	Convert harm into benefit
3	Local quality	23	Feedback
4	Asymmetry	24	Mediator
5	Consolidation	25	Self-service
6	Universality	26	Copying
7	Nesting (Matrioshka)	27	Dispose
8	Counterweight	28	Replacement of mechanical system
9	Prior counteraction	29	Pneumatic or hydraulic construction
10	Prior action	30	Flexible films or thin membranes
11	Cushion in advance	31	Porous materials
12	Equipotentiality	32	Changing the colour
13	Do it in reverse	33	Homogeneity
14	Spheroidality	34	Rejecting and regenerating parts
15	Dynamicity	35	Transformation of properties
16	Partial or excessive action	36	Phase transition
17	Transition into a new dimension	37	Thermal expansion
18	Mechanical vibration	38	Accelerated oxidation
19	Periodic action	39	Inert environment
20	Continuity of useful action	40	Composite materials

shuller [1997]) is counted. The principles with the highest occurrences are the favorite principles for improving the system's performance regarding that technical parameter. In addition to the results presented in Ivashkov and Souchkov [2004], we added two columns for the second and third best principles for each parameter. The result is shown in Table A.3.

For some technical parameters the second and/or third best principles are left blank. This indicates that the difference between the occurrence of the favorite and second or the second and third best principles was so large that the fitness of that IP can be questioned.

A.1.2 NEGATIVE PRIORITY MATRIX, PM^-

The negative priority matrix PM- is determined in the same manner. The only difference is that the columns, instead of the rows are analyzed. The resulting negative priority matrix is shown in Table A.4. This matrix can be used to reduce the negative consequences of a given technical parameter on a system.

Table A.3

The positive priority matrix PM^{+}, found from counting the times an innovative principle is mentioned in the row of technical parameter. The numbers in the columns marked 1st, 2nd, and 3rd are pointers to the relevant innovative principles; see Table A.2

	Technical Parameter	1st	2nd	3d
1	Weight of moving object	35		28
2	Weight of stationary object	35	10, 19, 28	1
3	Length of moving object	1, 29	15, 35	4
4	Length of stationary object	35	28	1, 10, 14, 26
5	Area of moving object	2, 15	13, 26, 30	4
6	Area of stationary object	18	2	35
7	Volume of moving object	1, 35	2, 10, 29	4, 15, 34
8	Volume of stationary object	35		
9	Speed	28	35	13
10	Force (intensity)	35	10, 18, 37	36
11	Stress or pressure	35	10	
12	Shape	10	1, 14, 15	32, 34
13	Stability of the object's composition	35		
14	Strength	3, 35	10	40
15	Duration of action of moving object	19	35	3, 10
16	Duration of action of stationary object	35	1, 10, 16	40
17	Temperature	35	19	
18	Illumination intensity	19	32	
19	Use of energy by moving object	35	19	
20	Use of energy by stationary object	19, 35	18, 27	many
21	Power	35	19	2, 10
22	Loss of energy	7	35	2
23	Loss of substance	10	35	
24	Loss of information	10	26, 35	
25	Loss of time	10	35	18
26	Quantity of substance/the matter	35		
27	Reliability	35	11	10
28	Measurement accuracy	32	28	
29	Manufacturing precision	32		
30	Object-affected harmful factors	22	35	
31	Object-generated harmful factors	22, 35	2	1, 39
32	Ease of manufacture	1		
33	Ease of operation	1	13	2
34	Ease of repair	1	10	
35	Adaptability or versatility	35		
36	Device complexity	13, 26	1, 28	2, 10
37	Difficulty of detecting and measuring	28	35	16, 18, 26, 27
38	Extent of automation	35	13	28
39	Productivity	10	35	

Table A.4

The negative priority matrix PM^-, found from counting the times an innovative principle is mentioned in the column of technical parameter. The numbers in the columns marked 1st, 2nd, and 3rd are pointers to the relevant innovative principles; see Table A.2

	Technical Parameter	1st	2nd	3d
1	Weight of moving object	35	8, 15, 28	2, 26, 40
2	Weight of stationary object	35	26, 27	
3	Length of moving object	1	15	14, 17, 28, 29
4	Length of stationary object	14, 26	1, 10, 35	7, 15, 28
5	Area of moving object	17	13, 15, 26	10, 29
6	Area of stationary object	16, 18, 35, 40	2, 10, 17, 39	30, 36
7	Volume of moving object	35	2, 10	29
8	Volume of stationary object	35		
9	Speed	35	28	
10	Force (intensity)	35	10	
11	Stress or pressure	35	10	
12	Shape	1, 35	29	10, 14, 15
13	Stability of the object's composition	35		
14	Strength	28	3	10
15	Duration of action of moving object	35	3	19
16	Duration of action of stationary object	16	10, 35	1
17	Temperature	35	19	
18	Illumination intensity	19	32	1, 13
19	Use of energy by moving object	35	19	
20	Use of energy by stationary object	1	35	18, 19
21	Power	35	10, 19	2, 18
22	Loss of energy	35	2	
23	Loss of substance	10	35	
24	Loss of information	10	24	
25	Loss of time	10	35	28
26	Quantity of substance/the matter	35	3	
27	Reliability	10	11	40
28	Measurement accuracy	28	32	
29	Manufacturing precision	32	10	28
30	Object-affected harmful factors	22	35	2
31	Object-generated harmful factors	35	2	
32	Ease of manufacture	1	35	
33	Ease of operation	1	32, 35	2, 28
34	Ease of repair	1	10	
35	Adaptability or versatility	15		
36	Device complexity	1	26	10
37	Difficulty of detecting and measuring	35		
38	Extent of automation	35	2	
39	Productivity	35		

B Types of Failure Modes and Effect Analysis

The history of FMEA goes back to the early 1940's when it was originated by the U.S. military and was further developed by the aerospace and automotive industries. Several formal FMEA standards have been developed for different types of industries based on the ISO9001 standards. The following types can be distinguished:

System FMEA: (SFMEA) is applied in interaction of parts of a systems, and used to analyze complete systems and/or sub-systems during the concept design stage. The focus is on minimizing failure effects on the System to provide System objectives like, quality, reliability, robustness, maintenance, life-cycle costs.

Design FMEA: (DFMEA) is applied to the development of a product and used to analyze a product design before it is released to the manufacturing stage. The focus is on minimizing failure effects on the Product to meet objectives like, quality, reliability, robustness, maintenance, life-cycle costs.

Process FMEA: (PFMEA) is applied to the process and used to analyze manufacturing and/or assembly processes. It focuses on minimizing failure effects on the Processes to provide maximize total Process objectives.

C Document Template

This appendix shows a simple general document template in reduced size. It does not show content, as that is treated in §4.13.2; it merely shows the frame in which a document can be produced, and the essential meta-information it should contain.

System Requirement Specification
Device XYZ

Company information/logo

Author	Distribution list:
Date	Manager SE
Document Status	. . .
Document ID	

Short summary:
This should be concise and list the main results and main requirements or design features of the (sub)system.

Applicable documents:
Controlling documents[1]*::*

Title	Doc ID	Version	Date

Controlled documents[2]*::*

Title	Doc ID	Version	Date

Referenced documents[3]*::*

Title	Doc ID	Version	Date

page 1

[1]What documents higher in the hierarchy influence this document?
[2]What documents lower in the hierarchy are influenced by this document?
[3]Other reference documents (literature, test reports, etc.).

Document history:

Version	Date	Who	Changes
Create			

Review team:
Author
Manager SE
. . .

page 2

Bibliography

Altshuller, G. S.: 1997; Forty Principles: TRIZ Keys to Technical Innovation; TRIZ Tools, vol. 1; Technical Innovation Center, Worcester, MA.

van Amerongen, J.: 2010; Dynamical Systems for Creative Technology; Controllab Products B.V., Enschede.

Arunski, K., J. Martin, P. Brown and D. Buede: 1999; Systems Engineering Overview; presentation to the Texas Board of Professional Engineers; September 1999; INCOSE.

Baumann, H. and A. Tillman: 2004; The Hitchhiker's guide to LCA: an Orientation in Life-Cycle Assessment Methodology and Application; Studentlitteratur AB. http://books.google.nl/books?id=3QdsQgAACAAJ

Blanchard, B. S. and W. J. Fabrycky: 2011; Systems Engineering and Analysis; Prentice Hall, Upper Saddle River, NJ. http://www.pearsoned.co.uk/bookshop/detail.asp?item=100000000374477

Blessing, L. T. M. and A. Chakrabarti: 2009; DRM: a Design Research Methodology; Springer. http://www.springer.com/engineering/mechanical+engineering/book/978-1-84882-586-4

Boardman, J. and B. Sauser: 2008; Systems Thinking: coping with 21st Century Problems, CRC Press, Boca Raton. http://www.crcnetbase.com/isbn/9781420054927

Boardman, J., B. Sauser, L. John and R. Edson: 2009; The conceptagon: A framework for systems thinking and systems practice; in Systems, Man and Cybernetics. IEEE International Conference on. http://dx.doi.org/10.1109/ICSMC.2009.5346211

Bolton, W.: 2012; Mechatronics: Electronic Control Systems in Mechanical and Electrical Engineering; Pearson, Harlow; 3rd edn.

Bonnema, G. M.: 2008; FunKey Architecting: An Integrated Approach to System Architecting Using Functions, Key Drivers and System Budgets; PhD-thesis; University of Twente. http://purl.org/utwente/58868

Bonnema, G. M.: 2011; Insight, innovation, and the big picture in system design; Systems Engineering; vol. 14 (3): pp. 223–238. http://dx.doi.org/10.1002/sys.20174

Bonnema, G. M.: 2014; Communication in multidisciplinary systems architecting; in 24th CIRP Design Conference, Milano.

Bonnema, G. M., P. D. Borches, R. G. Kauw-A-Tjoe and F. J. v. Houten: 2010; Communication: Key Factor in Multidisciplinary System Design; in 8th Annual Conference on Systems Engineering Research (CSER); Stevens Institute of Technology, Hoboken, NJ.

http://purl.utwente.nl/publications/71140

Borches, P. D. and G. M. Bonnema: 2010; A3 Architecture Overviews: Focusing architectural knowledge to support evolution of complex systems; in 20th Annual INCOSE International Symposium; Chicago. http://purl.utwente.nl/publications/77042

Borches Juzgado, P. D.: 2010; A3 Architecture Overviews: A Tool for Effective Communication in Product Evolution; PhD thesis; University of Twente. http://purl.utwente.nl/publications/75284

Braungart, M. and W. McDonough: 2009; Cradle to Cradle; Random House. http://books.google.nl/books?id=13hfHzBstcEC

Broenink, J. F.: 1990; Computer aided physical systems modeling and simulation: A bond-graph approach. Ph.D. thesis, University of Twente, Enschede, The Netherlands.

Clark, H. H. and S. E. Brennan: 1991; Grounding in communication. In Perspectives on Socially Shared Cognition; American Psychological Association, Washington, DC.

Conway, M. E.: 1968; How Do Committes Invent?; Datamation; vol. 14 (5): pp. 28–31. http://www.melconway.com/research/committees.html

Dailey, K. W.: 2004; The FMEA Pocket Handbook; DW Publishing Co.

Daniels, J. and T. Bahill: 2004; The hybrid process that combines traditional requirements and use cases; Systems Engineering, The Journal of the International Council on Systems Engineering; vol. 7 (4): pp. 303–319.

De Carvalho, M. A. and N. Back: 2000; Cross-Fertilization Between TRIZ and the Systematic Approach to Product Planning and Conceptual Design; in TRIZCON2000; Altshuller Institute for TRIZ studies, Nashua, NH.

van Dongen, L. A. M.: 2011; Maintenance Engineering: Instandhouding van Verbindingen; inaugural address University of Twente, Enschede, The Netherlands. http://www.utwente.nl/academischeplecthtigheden/oraties/archief/2007–2014/Oratieboekje˙van˙%20Dongen.pdf.

Eger, A., G. M. Bonnema, E. Lutters and M. C. v. d. Voort: 2013; Product Design; eleven international publishing. http://www.elevenpub.com/social-sciences/catalogus/product-design-1#

Eising, F.: 2007; Als je het begrijpt kun je het veranderen; Inaugural address University of Twente Enschede, The Netherlands. http://www.utwente.nl/academischeplechtigheden/oraties/archief/2007–2014/eising.pdf.

van Exel, N. J. A.: 2011; Behavioural Economic Perspectives on Inertia in Travel Decision Making; PhD Thesis; Vrije Universiteit Amerstdam. http://www.jobvanexel.nl/vanExel_full%20manuscript.pdf

Gadd, K.: 2011; Thinking in Time and Scale; pp. 69–93; John Wiley & Sons, Ltd. http://dx.doi.org/10.1002/9780470684320.ch4

Gelb, M. J.: 1998; How to Think Like Leonardo da Vinci: Seven Steps to Genius Everyday; Delacorte Press, New York.

Giancoli, D. C.: 2008; Physics for Scientists & Engineers; Pearson Education; 4th edn.

Gulatti, R. K. and S. D. Eppinger: 1996; The Coupling of Product Architecture and Organizational Structure Decisions; Working Paper 3906; MIT Sloan School of Management. http://web.mit.edu/eppinger/www/pdf/Gulati_SloanWP3906.pdf

Heemels, W., L. Somers, P. van den Bosch, Z. Yuan, B. van der Wijst, A. van den Brand and G. J. Muller: 2006; The key driver method; in Boderc: Model-Based Design of High-Tech Systems; pp. 27–42; Embedded Systems Institute, Eindhoven. http://repository.tue.nl/656087

INCOSE SEH Working Group: 2008; Systems Engineering Handbook; INCOSE; 3rd edn.

Ivashkov, M. and V. Souchkov: 2004; Establishing Priority of TRIZ Inventive Principles in Early Design; in Design 2004; Dubrovnik.

Kapurch, S. J. and N. E. Rainwater: 2007; NASA Systems Engineering Handbook SP/-2007-6105 Rev 1; NASA.
http://ntrs.nasa.gov/archive/nasa/casi.ntrs.nasa.gov/20080008301_2008008500.pdf

Kauw-A-Tjoe, R. G.: 2009; The Cleaning Robot Project; Master thesis; University of Twente.

Kossiakoff, A. and W. Sweet: 2003; Systems Engineering Principles and Practice; John Wiley & Sons.

MacKay, D. J. C.: 2008; Sustainable Energy without the Hot Air; UIT, Cambridge.
http://www.withouthotair.com/

Maier, M. W. and E. Rechtin: 2000; The Art of Systems Architecting; CRC Press, Boca Raton; 2nd edn.

Martin, J. N.: 2004; The Seven Samurai of Systems Engineering: Dealing with the Complexity of 7 Interrelated Systems; in Fourteenth Annual International Symposium of the International Council on Systems Engineering (INCOSE); INCOSE, Toulouse.
http://www.incose.org/wma/library/docs/Seven_Samurai-Martin-paper-v040316a.pdf

Martin, J. N. and H. L. Davidz: 2007; Systems Engineering Case Study Development; in 5th Annual Conference on Systems Engineering Research 2007 (CSER2007); Stevens Institute of Technology, Hoboken, New Jersey.
http://sse.stevens.edu/fileadmin/cser/2007/proceedings/9.pdf

Mascitelli, R.: 2011; Mastering Lean Product Development: A Practical, Event-Driven Process for Maximizing Speed, Profits, and Quality; Technology Perspectives, Northbridge, CA; 1st edn.

Mikulak, R. J., R. McDermott and M. Beauregard: 2008; The Basics of FMEA; Productivity Press; 2nd edn.

Muller, G.: 2011a; System Integration How-To.
http://www.gaudisite.nl/SystemIntegrationHowToPaper.pdf

Muller, G.: 2011b; Systems Architecting: A Business Perspective; CRC Press, Boca Raton, FL. http://books.google.com/books?id=MDxSSAAACAAJ

Muller, G.: 2013; Systems Engineering Research Methods; Procedia Computer Science; vol. 16 (0): pp. 1092–1101.
http://www.sciencedirect.com/science/article/pii/S1877050913001166

Muller, G. J.: 2004a; CAFCR: A Multi-view Method for Embedded Systems Architecting; PhD thesis; Delft University of Technology.
http://www.gaudisite.nl/Thesis.htmlhttp://www.gaudisite.nl/ThesisBook.pdf

Muller, G. J.: 2004b; The Waferstepper Challenge: Innovation and Reliability despite Complexity.
http://www.gaudisite.nl/IRCwaferstepperPaper.pdf

Muller, G. J.: 2007; Design objectives and design understandability; Last accessed on 13 May 2011.
http://www.gaudisite.nl/DesignUnderstandabilitySlides.pdf

Nonaka, I. and H. Takeuchi: 1995; The Knowledge-Creating Company : How Japanese Companies Create the Dynamics of Innovation; Oxford University Press.

Object Management Group: ; The Official OMG SysML site; Last accessed on 7 november 2013.
http://www.omgsysml.org

Onwubolu, G. C.: 2005; Mechatronics: Principles and Applications; Elsevier Butterworth-Heinemann, Amsterdam.

Richmond, B.: 1993; Systems thinking: Critical thinking skills for the 1990s and beyond; System Dynamics Review; vol. 9 (2): pp. 113–133.
http://dx.doi.org/10.1002/sdr.4260090203

Sage, A. P. and J. E. Armstrong jr.: 2000; Introduction to System Engineering; John Wiley and Sons.

Salamatov, Y. P.: 1999; TRIZ: the Right Solution at the Right Time: a Guide to Innovative Problem Solving; Insytec, Hattem, The Netherlands.

Schramm, W.: 1954; The Process and Effects of Mass Communication; University of Illinois Press, Urbana, IL.

Schön, D. and J. Bennet: 1996; Reflective Conversation with Materials; in Bringing Design to Software, edited by T. Winograd; pp. 171–184; ACM Press, New York. http://hci.stanford.edu/bds/9-schon.html

Suh, N. P.: 2005; Complexity in Engineering; Annals of CIRP; vol. 2 (2005): pp. 581–598.

Wikipedia: 2015a; Deepwater Horizon oil spill; retrieved 20150615.
http://en.wikipedia.org/wiki/Deepwater_Horizon_oil_spill

Wikipedia: 2015b; Piper Alpha; retrieved 20150615.
https://en.wikipedia.org/wiki/Piper_Alpha

Wiulsrød, B., G. Muller and M. Pennotti: 2011; Architecting Diesel Engine Control System using A3 Architecture Overview; in INCOSE International Symposium 2011; INCOSE, Rome.
http://www.gaudisite.nl/INCOSE2012_Wiulsr%C3%B8dEtAl_DieselEngineA3.pdf

Wojtkowski, W. and W. G. Wojtkowski: 2002; Storytelling: its role in information visualization; Fifth European Systems Science Congress, 16-19 October 2002; Association Française de Science des Systèmes Cybernétiques, Cognitifs et Techniques; Systems Science European Union, Crete, Greece. http://www.afscet.asso.fr/resSystemica/Crete02/Wojtkowski.pdf

Index

A

A3 Architecture Overview, *see* A3AO
A3AO, 63
  functional view, 63
  physical view, 63
  quantification view, 63
acceptance test, 18
accepted system, 18
agility, 37
architecture, 56, 85

B

balance, 85
baseline, 11, 87
budget, 58
  – items, 58

C

circuit breaker, 42
common ground, 37
consequences, 71
context diagram, 46
controversy of system design, 23
Conway's law, 37
cost, 42
coupling matrix, 61
cradle to cradle, 40
CSEP, 7
customer wish, 11

D

deadlock, 52
decision tree, 72
decomposition–composition thinking, 34
derived functions, 78
design, 23
development process, 10
differential equations, 82
documentation, 75
dynamic thinking, 25

E

effect analysis, 66
environment, 57
errors
  independent, 33
  systematic, 33

F

failure mode, 66
Failure Mode and Effect Analysis, *see* FMEA
feed-forward control, 29
feedback control, 27
feedback thinking, 27
FMEA, 66
frequency domain, 26
Function, 45, 50
function, 57
function tree, 50
functional block diagram, 51
functional model, 33

G

goal of system design, 8
guesstimation, 58
guideline, 7
  www.leidraadse.nl, 7

H

handling, 42
hierarchical thinking, 35

I

influence diagram, 53, 54
  logical, 53
  pictorial, 53
infrastructure, 45
integration, 16, 34
interface, 13, 57
  management, 55

ISO date format, 77
    short, 76, 77

K

Key driver, 60
key drivers, 85

L

life-cycle thinking, 38
    organization, 40
    product, 38
    resource, 40
life-cycle analysis, 40
long-lead items, 14, 79

M

margin, 59
    explicit, 59
    implicit, 59
milestone, 41

N

N^2 diagram, 55
    functional, 56
    modular, 56
nine-window diagram, 45

O

open issues, 81
operational thinking, 30
organizational thinking, 37

P

planning, 42
probability, 71
problem domain, 9
process, 1
product life-cycle, 38
program, 42
project life-cycle, 38

Q

question generator, 83

R

report investigation, 79
    test, 79
    test protocol, 79
resource life-cycle, 38
resources, 40
reviewing, 80
    register, 74
risk priority number, *see* RPN
risk thinking, 42
RPN, 70

S

safety thinking, 41
scales thinking, 31
scenario, 30, 48
    story, 30
    storyboard, 30
scientific thinking, 32
solution domain, 9
specific-generic thinking, 29
spiral development, 53
Stakeholder, 45
standard, 14
state, 52
state transition diagram, 52
story telling, 48
subsystem, 13, 57
subsystem requirements, 15
system, 7
    architecture, 37, 56
    design, 13
system design specification, 78
system of systems, 45
system requirement specification, 12, 77
system requirements, 12
system specifications, 12
system under design, 8
    S.U.D., 8
systems engineer, 2
systems engineering, 1
systems thinking, 23

T

technical, 42
template, 77
Theory of Inventive Problem Solving, *see* TRIZ
thinking tracks, 2
time boxing, 89
time domain, 26
time scales, 26
tools, 2
top-level design, 13
trade-offs, 12, 85
transition, 52
TRIZ, 45, 93
  contradiction matrix, 62

U

uncertain information, 23
use case, 49
Utility, 45

V

validation, 17, 86
vee model, 18
verification, 16, 17

W

WWHWWW, 83

9781498751261